ELEMENTS
OF
STEEL
DESIGN

ELEMENTS OF STEEL DESIGN

H. W. Morrow

Professor
Nassau Community College
Garden City, New York

PRENTICE-HALL, INC., Englewood Cliffs, N.J. 07632

Library of Congress Cataloging-in-Publication Data

MORROW, H. W. (HAROLD W.)
  Elements of steel design.
  Includes index.
  1. Building, Iron and steel. 2. Structural design. I. Title.
TA684.M714 1987   624.1′821   86-9419
ISBN 0-13-272717-X

Editorial/production supervision: Colleen Brosnan
Cover design: Wanda Lubelska Design
Manufacturing buyer: Rhett Conklin

Printed in the United States of America
10 9 8 7 6 5 4 3 2 1

ISBN 0-13-272717-X 025

Prentice-Hall International (UK) Limited, *London*
Prentice-Hall of Australia Pty. Limited, *Sydney*
Prentice-Hall Canada Inc., *Toronto*
Prentice-Hall Hispanoamericana, S.A., *Mexico*
Prentice-Hall of India Private Limited, *New Delhi*
Prentice-Hall of Japan, Inc., *Tokyo*
Prentice-Hall of Southeast Asia Pte. Ltd., *Singapore*
Editora Prentice-Hall do Brasil, Ltda., *Rio de Janeiro*

To my mother and father,

Hazel Nicholson Morrow
Harold S. Morrow

CONTENTS

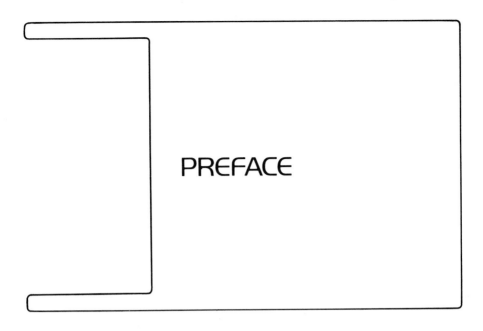

PREFACE

This book is written to provide a clear discussion of the principles of structural steel design at an elementary level. It is appropriate for students enrolled in an engineering or technology curricula such as civil, construction, and construction management and for steel design courses for architects. It can also be used for self-study by engineers, architects, and technicians. It may also be used as a reference for courses in advanced steel design. The book assumes that the reader has had an introductory course in statics and strength of materials.

The topics covered in the text have been limited to those thought necessary for an introductory course. Since the book is intended for use at an elementary level, care has been taken to introduce one concept at a time and to reinforce each concept with examples. The examples are presented in small steps to ensure a smooth transition to the topics that follow. Complex theory has been minimized and over ninety example problems are solved in detail. The objective has been the understanding of the structural behavior of each of the members that make up a structure, the application of the relevant specifications, and finally, the design of the member.

More than two hundred exercise problems on various levels of difficulty are provided. In addition, most of the examples and exercise problems have been formulated so that alternative data can be used if desired. The book can be used for several years without repeating assignments and can be used for classes of varying abilities.

The book has been coordinated with the American Institute of Steel, *Construction Specifications for the Design, Fabrication, and Erection of Structural*

Steel for Buildings (8th ed., 1978). The *Steel Construction Manual* should be available for use with the text.

Chapter 1 begins with an introduction to steel structures. It covers such topics as framed structures and their parts; properties and types of structural steel and their shapes; allowable stresses; factors of safety; codes and specifications; and classification of members. In Chapter 2 we consider the various possible loads on structures.

Chapters 3 through 6 are concerned with the detailed design of various members of a structure. They are tension members, columns, beams, and beam-columns. Each chapter begins with a discussion of the structural behavior of the member, a discussion of the relevant specification, and concludes with the design of the member.

Chapters 7 and 8 on bolted and welded connections follow a parallel development. The various types of bolted and welded connections can be studied together if desired. The chapters begin with a discussion of bolts or welds, and continue with axially loaded and eccentrically loaded connections and connections subject to shear and tension. The chapters conclude with such simply supported building frame connections as shear and seated beam connections, and moment connections with flange and vertical web plates and end plates. More material on connections is included than is usual in an introductory text. However, no text can adequately cover all of the possible structural connections.

No reference is made in the text to SI units. Use of data converted to SI units by the author would serve no useful purpose and only lead to confusion on the part of the reader. This is consistent with the continued use of common units in the AISC *Specifications* and *Manual* and in the steel industry.

Reviewers and users of the text are encouraged to write the author with comments concerning the book. He is especially interested in identifying errors and in suggestions for adding or deleting topics in a possible future edition.

The text was written during a sabbatical leave in 1984–85. Copies of the manuscript were used at Nassau Community College during the school year 1985–86. In this regard, acknowledgment is made to the assistance of Professor Andrew C. Kowalik in arranging for classroom testing of the manuscript.

Finally, special thanks are due my good friend, George J. Moses, for his encouragement during the preparation of the manuscript and the production of this book.

Harold W. Morrow

New York, NY
Ft. Lauderdale, FL

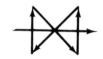

INTRODUCTION TO STEEL STRUCTURES

1-1 INTRODUCTION

Design requires the development of plans that enable the structure to fulfill its intended use. Called *functional design* or planning, this phase of design is concerned with such decisions as the material to be used, the number of stories and location of columns in a building, and the width and location of a bridge. Cost is an important factor in these decisions. Design also requires the determination of the size and proportions of the members that form the structure. The members must be able to safely resist the forces produced by the loads on the structure. This phase of design is called *detail design*. In detail design the principles of mechanics and strength of materials together with the appropriate specifications are used to ensure that the various members of the structure have adequate strength and stiffness.

The designer must first learn to design the individual members before designing the complete structure. In this book we study the detailed design of various structural members and their connections. Tension members, columns, beams, and beam-columns are covered in Chapters 3 through 6 and bolted and welded connections in Chapters 7 and 8.

1-2 FRAMED STRUCTURES AND THEIR PARTS

Although we are concerned primarily with the design of the individual parts of a structure, we begin with an overview of various types of framed structures and their parts.

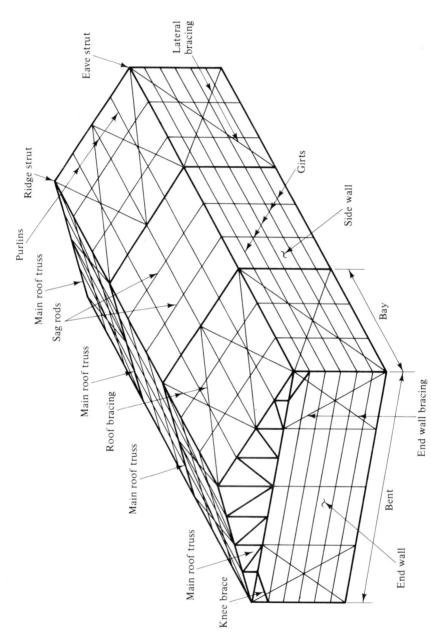

Figure 1-1 Truss on column framing for an industrial building

Eave strut

Lateral bracing

Ridge strut

Purlins

Main roof truss

Sag rods

Main roof truss

Roof bracing

Main roof truss

Main roof truss

Knee brace

Girts

Side wall

Bay

End wall bracing

Bent

End wall

A building structure consisting of a steel framework or skeleton is made up of tension members, columns, beams, and members under combined bending and axial loads. The members are joined together by bolted and/or welded connections or joints. The members must resist the action of the weight of the building and its contents, snow and ice on the roof, the pressure and suction of the wind, earthquake forces, and impact. All of the loads must be transmitted by the structural members to the foundation and then to the ground.

Commercial and industrial buildings are framed in a number of different ways. Commonly they are one story high with roofs supported by trusses, beams, or open-web joists. The trusses, beams, or open-web joists may be supported by walls or columns.

Truss-on-column framing for an industrial building is illustrated in Fig. 1-1. The truss and supporting columns together are called a *bent*. The spaces between the bents are called *bays*. The roof is supported by the purlins, which in turn are supported by the trusses at their panel points. The side and end walls are supported by the columns and girts. Horizontal forces from the wind, an earthquake, or impact acting in a transverse direction (perpendicular to the sides of the building) are resisted by the side-wall girts, which transmit their loads to the bents, where they are resisted by the knee braces and end-wall bracing. Horizontal forces in a longitudinal direction (perpendicular to the end walls) are resisted at the roof by purlins and roof bracing and below the roof by end-wall girts and side-wall lateral bracing.

Detail shown in Fig. 1-3

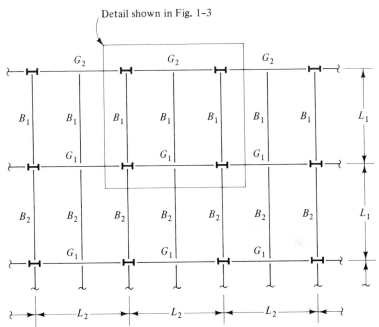

Figure 1-2   Floor framing plan for a multistory building.

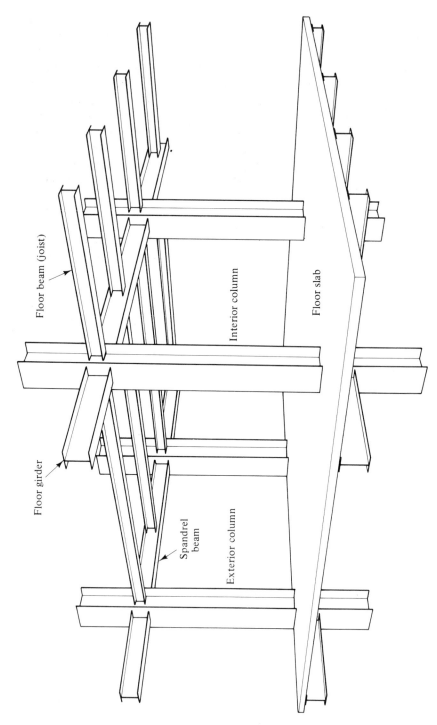

Figure 1-3 Detail of framing for the multistory building shown in Fig. 1-2.

Floor beam (joist)

Floor girder

Spandrel beam

Exterior column

Interior column

Floor slab

A multistory building frame usually consists of girders, beams, and columns. The members are joined by rigid, semirigid, or simply supported connections. If the connections are not rigid, bracing must be provided. The plan view of a multistory building is shown in Fig. 1-2, and a detail of the building with *simply supported* connections is shown in Fig. 1-3.

In the multistory building shown the floor loads are transmitted from the floor slab to the beams. The beams are supported at their ends by either columns or girders. The ends of the girders are also supported by the columns. The loads are transmitted between floors by the columns. For tall multistory buildings the horizontal forces must be resisted by rigid connections between the beams or girders and the columns or by various types of bracing.

Moment-resisting or rigid connections and various kinds of bracing for multistory building frames are illustrated in Fig. 1-4. Diagonal bracing that

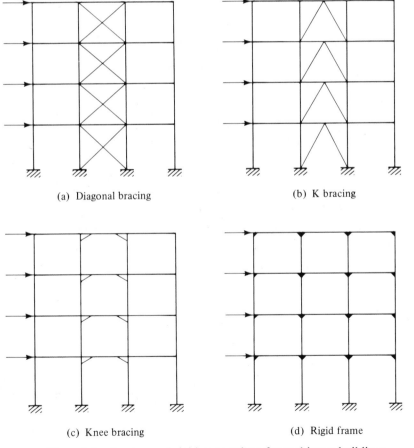

(a) Diagonal bracing   (b) K bracing

(c) Knee bracing   (d) Rigid frame

Figure 1-4   Bracing and rigid connections for multistory buildings.

forms vertical trusses with the columns and beams is the most efficient but blocks the opening between the beams and columns and can be used only where no interference occurs [Fig. 1-4(a)]. The K bracing system is also efficient but blocks a considerable part of the opening [Fig. 1-4(b)]. Modified K bracing, where the braces are separated at the top, is efficient but also blocks

Figure 1-5 Philadelphia Life Building. (Courtesy of Bethlehem Steel Corporation.)

a considerable part of the opening. Knee bracing is less efficient but leaves most of the opening free [Fig. 1-4(c)]. Frames made with rigid connections have openings that are completely free [Fig. 1-4(d)]. However, rigid frames require large member sizes on the lower floors to limit horizontal deflections and may be uneconomical for high-rise buildings over about 40 stories.

Moment-resisting connections and bracing may be combined in the same structure. The 17-story steel frame of the Philadelphia Life Building shown in Fig. 1-5 is an example of this combination. Moment connections in the building frame resist longitudinal wind loads, while braced frames resist the transverse loads. The connections are welded or field bolted with ASTM A325 high-strength bolts.

To avoid long-span trusses and interior columns, most designers use either suspension cables to support the roof or a shell-type structure for the roof. An interesting exception to this practice is the mainly framed structure shown in Fig. 1-6. The figure shows most of the eight towers and the four trusses of the Byrne Meadowlands Arena in East Rutherford, New Jersey. In the arena, four box trusses arch upward 35 ft above their tower supports. The trusses cross at four points to form in plan view a figure consisting of two vertical lines crossing two horizontal lines (as in the game of tic-tac-toe). The eight steel-frame towers that support the ends of the trusses rise 95 ft above the arena floor. The arena encloses an area of over $150,000 \text{ ft}^2$.

Figure 1-6 Byrne Meadowlands Arena. (Courtesy of Bethlehem Steel Corporation.)

1-3 IRON-CARBON ALLOY

The terminology for a combination or alloy of iron and carbon is somewhat confusing. The term *iron* is applied to both the very low carbon material, cast iron, and the very high carbon material, wrought iron. In many ways steels are intermediate in carbon content between cast iron and wrought iron. Approximate limits for carbon in steel are between 0.04 and 2.25 percent. (The limits for structural steel are between 0.15 and 1.7 percent.)

The use of an iron–carbon alloy as a structural material has gone through three periods as a result of advances in metal technology. Although considerable overlap occurred between the periods, approximate dates are as follows: use of cast iron from 1779 to 1840, wrought iron from 1840 to 1890, and steel after 1890.

Cast Iron

The first use of cast iron as a structural material was on a 100-ft span bridge over the Severn River at Coolbrookdale, England. The bridge was completed in 1779 and still stands today. Additional cast-iron bridges were built between 1780 and 1820. The use of cast iron in bridges was brief, however, because of failures due to brittle fracture in tension and the availability of wrought-iron shapes in commercial quantities by about 1840.

Wrought Iron

Wrought iron replaced cast iron because of its ability to permit large deformations without fracture. Because of that ability, wrought iron could be formed into plates. The plates were then cut and shaped into structural members. An early example of the use of wrought iron was the Brittania Bridge across the Menai Straits in Wales, which was completed in 1850. Made from wrought-iron plates and angles, it consisted of four spans—two 460-ft center spans and two 230-ft end spans.

Steel

With the development of the Bessemer process in 1856 and the open-hearth steel-making furnace in 1864, relatively large quantities of steel became available for the first time. Rolled bars and I-shapes were available by the 1870s. As early as 1874 steel was used on the Eads Bridge across the Mississippi River at St. Louis. The bridge had three spans—a center span of 520 ft and two end spans of 502 ft. Since about 1890 steel has replaced wrought iron in the construction industry.

1-4 STEEL AS A STRUCTURAL MATERIAL

The wide use of structural steel in the United States and the world would be obvious to even the most casual observer. With more attention, the observer would notice steel in buildings, bridges, transmission and water towers, and many other structures. The wide use of steel occurs because of its versatility, desirable properties, abundance, and competitive cost.

Steel is probably the most *versatile* commonly used structural material. Not only is its versatility apparent in the great variety of structures for which it is used but also in the many different forms possible in a single building structure or a complex of structures. This fact is especially evident in the steel frames of the Hughes Justice Complex in Trenton, New Jersey, shown in Fig. 1-7. The two frames consist of a nine-story office building and an elevated

Figure 1-7   Richard J. Hughes Justice Complex. (Courtesy of Bethlehem Steel Corporation.)

cube-shaped court building. The plan view of the office building looks like a wide inverted V. (The right-hand side of the inverted V is partially obscured in the figure.) The four-story court building is elevated by four massive 42-ft supporting columns and cantilevered out from the columns for 40 ft in two directions by steel trusses. The cube-shaped court building extends from the V's notch, which includes the public atrium and adjacent central forum. The complex also includes a two-level parking garage below grade.

Steel is also the principal structural metal in use today. A brief description of some of the *desirable properties* of structural steel are as follows:

1. *Elasticity.* Under load, steel follows Hooke's law up to high values of stress in both tension and compression. The behavior of steel can be predicted quite accurately. Simple linear or elastic theory based on Hooke's law can be used to design structural elements.

2. *Ductility.* The property of steel that permits it to undergo large deformations without fracture is called ductility. It gives steel the ability to resist sudden collapse and may be the single most important property of steel. Steel is a very "forgiving" material. Many of the simplifying assumptions used in structural steel design can be justified because of the ductility of steel. An example of this is the neglect of stress concentrations at bolt holes in the design of tension members under static loads. Additional examples will be given in the various chapters on member design.

3. *Strength and weight.* Steel has high strength per unit weight. The weight of the steel constitutes a small part of the load that can be supported by a steel structure. It permits more volume or space in the structure because the structural members are relatively slender. This property is important in the design of structures such as tall buildings, long-span bridges, and airplane hangars.

4. *Uniformity.* Because of the control exercised by the steel manufacturers, both the properties of steel and uniformity of structural shapes can be assured. This eliminates the need to overdesign a member because of uncertainty about the steel.

5. *Other advantages.* Some other advantages of structural steel are its permanence if properly maintained, ease of fabrication, speed of erection, possible reuse at a new location, and scrap value.

Steel also has *undesirable properties.* Some of these are as follows:

1. *Maintenance.* Corrosion is the enemy of a steel structure. Exposure to water and air has a devastating effect. Steel must be maintained by painting on a periodic basis, and drainage from the structure must be maintained. Some steel—called *weathering steel*—oxidizes when exposed to air or water. The steel forms a protective layer that resists additional corrosion. The protective layer has an attractive red-brown color and painting is not required.

2. *Fireproofing.* When the contents of a building burn, the heat raises the temperature of the structural steel in the building unless the steel is covered by insulating material. At high temperatures the strength of steel is drastically reduced. Steel at 1000°F has about 65 percent of the strength at room temperatures and at 1600°F about 15 percent. With such reduc-

tions in strength, buildings may collapse or members may undergo such large distortions that they must be removed and replaced.

3. *Buckling.* The high strength per unit weight for steel may not be an advantage in the design of slender compression members since they fail due to buckling rather than to a lack of material strength. Additional steel must be used to stiffen the member and prevent buckling.

4. *Fatigue.* A structural member subject to many stress reversals or even large changes in either tension or compression may fracture due to fatigue. Where fatigue may occur the member must be redesigned or the estimated strength of the member reduced.

1-5 PROPERTIES OF STRUCTURAL STEEL

Many of the properties of structural steel of interest to the designer can be described by the behavior of steel during a simple tension test. Standard procedures for this test are covered in a publication of the American Society for Testing and Materials (ASTM). In the test a steel specimen is gradually stretched or deformed, thus applying load to the specimen. Simultaneous readings of the load P and deformation Δ are taken at specified intervals. The stress f is found by dividing the load P by the original cross-sectional area A of the specimen ($f = P/A$), and the corresponding strain ϵ is found by dividing the deformation Δ by the original (gage) length L of the specimen ($\epsilon = \Delta/L$). The values obtained can be used to plot a stress–strain diagram. A typical *stress-strain diagram* for structural steel is shown in Fig. 1-8(a). The initial portion of that diagram with a larger horizontal scale is shown in Fig. 1-8(b).

The following properties of interest will be described by reference to Fig. 1-8.

1. *Modulus of elasticity.* The ratio of stress to strain or the slope of the initial straight-line part of the stress–strain curve is called the modulus of elasticity E. It is a measure of the stiffness of a material. The deflection of beams and the critical or buckling load for a slender column depends on stiffness. The modulus of elasticity for all structural steel is nearly constant and is commonly taken in structural steel design to be 29,000 ksi. The stiffness of steel is approximately three times larger than it is for aluminum.

2. *Proportional limit.* The maximum stress for which stress is proportional to strain or the maximum stress on the initial straight-line part of the stress–strain curve is called the proportional limit.

3. *Yield point.* The stress for which the strain increases without an increase in stress or the height of the horizontal portion of the stress–strain curve just beyond the proportional limit is called the yield point.

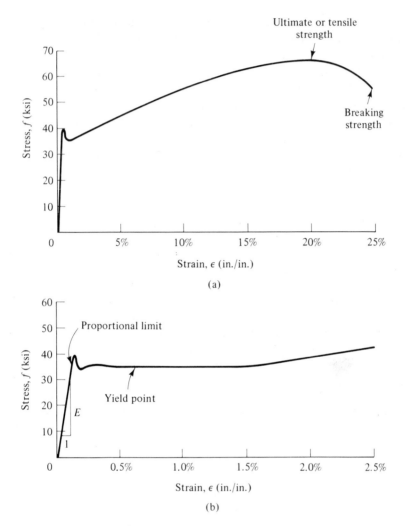

Figure 1-8 Typical stress–strain curve for ASTM A36 steel: (a) entire curve; (b) initial curve.

4. *Tensile strength.* The maximum stress on the stress–strain curve is called the tensile strength or ultimate strength.

5. *Ductility.* The property of steel that permits it to undergo large deformations without fracture is called ductility. It gives steel the ability to resist sudden collapse. Two measures of ductility are commonly used: reduction in area and elongation expressed as a percentage of initial length. The percentage elongation for the steel shown in Fig. 1-8(a) would be the maximum strain, 24.5 percent. The original (gage) length should be specified.

6. *Toughness.* Toughness is the ability of the material to absorb energy in large quantities. The *modulus of toughness* is the work done on a unit volume of material from zero force up to the force at the breaking point. It is equal to the area under the entire stress–strain curve.

7. *Yield strength.* Not all types of steel exhibit a well-defined yield point. For such steel, yield strength is defined as the stress that will cause the material to have a certain amount of permanent strain after unloading. The permanent strain ϵ_1 is commonly either 0.1 or 0.2 percent. The yield strength at a specified permanent strain is the stress at the intersection of the stress–strain curve and a line parallel to the modulus line through the specified permanent strain. The yield strength at a permanent strain ϵ_1 of 0.2 percent for the steel illustrated in Fig. 1-9 is 100 ksi, as shown.

8. *Yield stress.* This term is used to indicate the yield point for those steels that have a yield point or the yield strength for those steels that do not have a yield point.

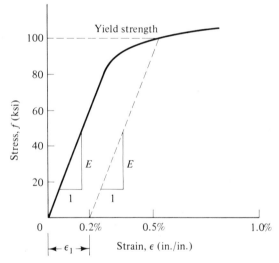

Figure 1-9 Typical yield strength at 0.2 percent strain for ASTM A514 steel.

1-6 TYPES OF STRUCTURAL STEEL

The essential elements in steel are metallic iron and the element nonmetallic carbon. Other elements, such as silicon, nickel, manganese, chromium, and copper, may be present in small quantities. When steel has significant amounts of some of these elements, it is an alloy. Although steel is almost entirely made up of iron and the elements added to iron are present in small quantities, they have a pronounced effect on the properties of steel. (Steel is usually more than 98 percent iron.)

Structural steel may be classified in accordance with the American Society for Testing and Materials (ASTM) A6 standard as carbon steel, high-strength low-alloy steel, and quenched and tempered alloy steel. Properties and product availability for various steels are given in Table 1-1.

TABLE 1-1   Structural Steel Properties and Plate and Shape Availability

Type of Steel	ASTM Designation	Minimum Yield Stress, F_y (ksi)	Tensile Strength, F_u (ksi)	Plate and Bar Thickness[a] (in.)	Shape Availability (Group)[b]
Carbon	A36	36	58–80	To 8	All
High-	A441	40	60	4–8	—
strength		42	63	$1\frac{1}{2}$–4	4 and 5
low-alloy		46	67	$\frac{3}{4}$–$1\frac{1}{2}$	3
		50	70	To $\frac{3}{4}$	1 and 2
	A572				
	Grade 42	42	60	To 6	All
	Grade 50	50	65	To 2	All
	Grade 60	60	75	To $1\frac{1}{4}$	1 and 2
	Grade 65	65	80	To $1\frac{1}{4}$	1
Corrosion-	A242	42	63	$1\frac{1}{2}$–4	4 and 5
resistant		46	67	$\frac{3}{4}$–$1\frac{1}{2}$	3
high-		50	70	To $\frac{3}{4}$	1 and 2
strength					
low-alloy	A588	42	63	5–8	—
		46	67	4–5	—
		50	70	To 4	All
Quenched	A514	90	100–130	$2\frac{1}{2}$–6	—
and		100	110–130	To $2\frac{1}{2}$	—
tempered					
alloy					

[a]Thicknesses do not include smaller dimension but do include larger dimension.
[b]See AISC *Manual*, Table 2, p. 1-6, for structural shape in each ASTM A6 group.
Source: Adapted from AISC *Manual*, Table 1, p. 1-5.

Carbon Steels

Carbon steels have specified maximum percentages of the elements carbon, manganese, silicon, and copper. Their strength is dependent on carbon and manganese. They are divided into four categories—low, mild, medium, and high—based on the carbon content. Mild carbon steel has a carbon content between 0.15 and 0.29 percent. The most commonly used structural steel, ASTM A36, has a carbon content from 0.25 to 0.29 percent, has a marked yield point at 36 ksi, and may be riveted, bolted, and welded. Increased carbon raises the yield point but reduces ductility and weldability.

High-strength Low-alloy Steels

These steels have higher yield stresses than carbon steels, ranging from 40 to 70 ksi. In addition to carbon and manganese, small amounts of such alloy elements as chromium, columbium, copper, nickel, vanadium, and others are added to increase the strength and corrosion resistance of the steel. High-strength low-alloy steels ASTM A441 and A514 may be economical where light members can be used and corrosion resistance reduces maintenance. They are rolled without heat treating.

Heat-treated Alloy Steels

Alloy steels are heat-treated by quenching and tempering to produce strong and tough steels. They have yield strengths from 80 to 110 ksi. Since they do not exhibit a well-defined yield point, the yield strength is defined by permanent strain. (For steel it is usually defined at 0.2 percent permanent strain.) The heat treating consists of quenching and then tempering the steel. Quenching is the rapid cooling of the steel with water or oil from temperatures of about 1650°F to about 350°F. Tempering is the reheating of the steel to about 1150°F, then allowing it to cool to room temperature. Although tempering steel reduces strength and hardness, it increases toughness and ductility. One heat-treated alloy steel—ASTM A514 is shown in Table 1-1.

1-7 STRUCTURAL STEEL SHAPES

Structural steel is rolled into a variety of shapes and sizes. The shapes are designated by the shape and size of their cross section. The *Manual of Steel Construction* published by the American Institute of Steel Construction (AISC) gives information on dimensions and sectional properties of the various steel shapes.

Commonly used structural shapes are shown in Fig. 1-10 and described in the following paragraphs.

1. *Wide-flange beams.* A wide-flange section has a cross section similar to an I. The two parallel horizontal parts of the cross section are called flanges and the vertical part is called the web. *Example:* W18 × 71 has a wide-flange beam shape approximately 18 in. deep and weighs 71 lb/ft.

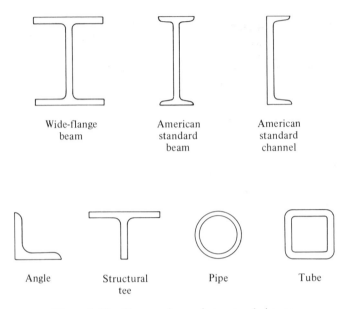

Wide-flange beam   American standard beam   American standard channel

Angle   Structural tee   Pipe   Tube

Figure 1-10   Commonly used structural shapes.

2. *American standard beam.* This section is commonly called an *I-beam.* The flanges are narrower than those of the wide-flange beam and the inner flange surfaces have a slope of approximately 9.2° (2 to 12 in.). *Example:* S20 × 66 has an American standard beam shape 20 in. deep and weighs 66 lb/ft.

3. *Bearing pile section.* Bearing pile sections are similar to the wide-flange beam except that the web and flange thickness are equal. *Example:* HP14 × 73 has a bearing pile shape approximately 14 in. deep and weighs 73 lb/ft.

4. *Miscellaneous shapes.* These sections have a doubly symmetric H-shape that cannot be classified as a W, S, or HP section. *Example:* M20 × 9 has a miscellaneous shape, is approximately 20 in. deep, and weighs 9 lb/ft.

5. *American standard channel.* This section is similar to the American standard beam with the flanges removed from one side. *Example:* C9 × 20 has an American standard channel shape, is 9 in. deep, and weighs 20 lb/ft.

6. *Angles.* Angles are shaped like an L. The horizontal and vertical parts are called *legs.* The inner and outer surfaces of each leg are parallel. The section may have either equal or unequal legs. *Examples:* (Equal-leg angle) L6 × 6 × $\frac{1}{2}$ has an angle shape L, both legs 6 in. long, and leg thickness $\frac{1}{2}$ in.; (unequal-leg angle) L5 × 3 × $\frac{5}{8}$ has an angle shape L, long leg 5 in., short leg 3 in., and leg thickness $\frac{5}{8}$ in.

7. *Structural tees.* A standard tee is made by cutting a wide-flange, miscellaneous, or an American standard section in half along a horizontal axis. The designation is similar to the W, M, or S shape. *Example:* (from W-shape) WT7 × 41; (from M-shape) MT7 × 9; (from S-shape) ST12 × 40.

8. *Pipes and tubes.* Various pipes and structural tubes, both square and rectangular, are manufactured from ASME A53, A500, and A501 carbon steel (see the AISC *Manual,* Table 4, p. 1-11).

1-8 ALLOWABLE STRESSES: FACTOR OF SAFETY

The allowable stress is the maximum stress that is considered safe for a material to support under certain loading conditions. It is the stress used in the design of load-supporting members of a structure. The allowable stress values are determined by tests on the material with various loading conditions and by experience gained from the performance of previous designs under service conditions. Allowable stresses are usually specified by codes and specifications.

The allowable stress in allowable stress design can be determined from a *factor of safety.* The safety factor may be defined as the ratio of some stress that represents the strength or buckling stress of a member to the allowable stress for the member. If based on the yield stress, the factor of safety is defined as the ratio of yield stress to the allowable stress:

$$\text{F.S.} = \frac{F_y}{F_{\text{all}}}$$

Similarly, the factor of safety based on the critical or buckling stress for a column is defined as the ratio of the critical stress to the allowable stress:

$$\text{F.S.} = \frac{F_{\text{cr}}}{F_{\text{all}}}$$

Since the allowable stresses are substantially lower than the yield stress or critical stress, factors of safety are numbers greater than 1. For the ductile failure of a tension member, the American Institute of Steel Construction (AISC) *Specifications* require a safety factor of 1.67. For a buckling failure in column design, factors of safety varying from 1.67 to 1.92 are recommended.

1-9 CODES AND SPECIFICATIONS

Most states and cities require structural design in their jurisdiction to conform to a building code. Some cities have their own codes; others use regional or national codes. Building codes generally cover all aspects of building safety, such as structural design, architectural details, elevator design, fireproofing, heating, ventilating, air conditioning, lighting, and plumbing. Many governmental bodies have incorporated the American Institute of Steel Construction (AISC) *Specifications* for structural steel design into their building code. The following is a partial list of *design codes* and *specifications* related to structural steel.

Design Codes

Basic Building Code (formerly BOCA), Building Officials and Code Administrators International, Chicago, Illinois. This code is used widely in the north-central and eastern states.

Uniform Building Code (ICBO), International Conference of Building Officials, Pasadena, California. This code is used widely in the western states.

National Building Code (National), the American Insurance Association, New York, New York. This code is used in various localities throughout the United States.

Southern Standard Building Code (SSBC), the Southern Building Code Congress, Birmingham, Alabama. This code is used widely in the southeast and south.

Specifications

American Institute of Steel Construction (AISC), *Specifications for the Design, Fabrication and Erection of Structural Steel for Buildings*, 8th ed., Chicago, 1978.

American Iron and Steel Institute (AISI), *Specifications for the Design of Cold-formed Steel Structural Members*, Washington, D.C., 1980.

American Welding Society (AWS), *Structural Welding Code*, Steel AWS D1.1-84, 9th ed., Miami, 1985.

American National Standards Institute (ANSI), *Minimum Design Loads in Buildings and Other Structures*, ANSI A58.1, New York.

American Association of State Highway and Transportation Officials (AASHTO), *Specifications for Highway Bridges*, 12th ed., Washington, D.C., 1977 (with 1978, 1979, 1980, and 1981 Interim Bridge Specifications).

American Railway Engineering Association (AREA), *Specifications for Steel Railway Bridges*, Chicago, 1980.

American Society for Testing and Materials (ASTM), Philadelphia. Publishes a series of volumes containing test methods, classifications, specifications, and so on. Material on structural steel is included.

1-10 CLASSIFICATION OF MEMBERS

For the purposes of design, structural members may be classified according to the type of loads they support. The classifications are as follows:

1. *Tension members:* members having axial tensile loads only.
2. *Compression members or columns:* members having only axial compressive loads.
3. *Bending members of beams:* members having only transverse loads.
4. *Combined bending and compression members or beam-columns:* members having both transverse loads and axial compressive loads. (Members may also have a combination of transverse loads and axial tensile loads.)

Structural steel is rolled in a variety of shapes and sizes to satisfy the design requirements of the various types of members. For a description, see Sec. 1-6.

1-11 ACCURACY OF CALCULATIONS

In the analysis of an existing member or the design of a member, the calculations will most likely be performed by an electronic calculator. Because of the large number of digits displayed by the calculator, the question arises as to how many digits should be retained in the intermediate steps of the calculations and in the final answer. Care should be taken to retain sufficient digits in the intermediate steps to ensure the required accuracy of the final answer. Hard-and-fast rules are difficult to formulate, but design data are rarely known to an accuracy greater than 0.2 percent. To maintain this accuracy we can use the following practical rule: use four digits to record numbers beginning with 1 and three digits to record numbers beginning with 2 through 9 from intermediate calculations. In view of the approximations made in steel design—starting with the assumed design loads—the final answer should only rarely be recorded to more than two or three significant digits.

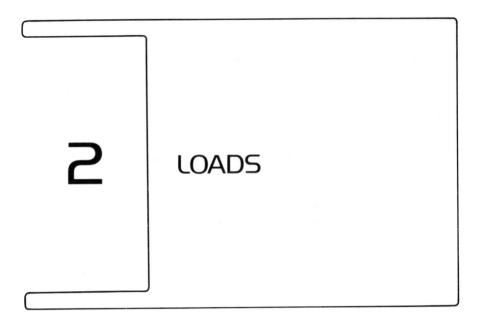

2 LOADS

2-1 INTRODUCTION

The structural designer begins with the determination of all the loads and forces that a structure or structural element must support. This includes all the forces that are caused by gravity, structural use, and the environment. Although some of the loads may be determined quite accurately, others must be based on current practice as reflected in the appropriate building code.

Loads may be classified as dead loads, live loads, snow loads, wind loads, impact, and earthquake forces. Each of the load classifications is discussed in the following sections.

2-2 DEAD LOADS

Loads that are always on the structure and are due to gravity forces are called *dead loads.* They include the weight of the structure and any components that are attached to the structure. Examples of dead loads in buildings are the weight of the floors, beams, columns, walls, plumbing, light fixtures, and so on. The weight of the structure itself can be determined accurately but only *after* each member has been designed.

2-3 LIVE LOADS

All loads other than dead loads are called *live loads.* They are caused by the occupancy and use of the building. Examples of occupancy loads are people, furniture, machinery, equipment, and stored material. Building live loads are usually regarded as uniformly distributed. Live loads cannot be determined as accurately as they are for dead loads. Minimum live-load requirements vary widely among the various building codes. Some representative values for floor live loads are given in Table 2-1. For an actual design, the designer should use values from the governing building code.

TABLE 2-1   Typical Building Design Live Loads

Type of Building	Live Load (psf)
Apartment house	
Corridors	80
Apartments	40
Public rooms	100
Assembly halls	
Fixed seats	60
Movable seats	100
Office buildings	
Offices	50
Lobbies	100
Corridors	100
Schools	
Classrooms	40
Corridors	100
Storage warehouse	
Light	125
Heavy	250

In the absence of a governing code the *National Standard Building Code* of the American National Standard Institute (ANSI) is usually followed. (See Sec. 1-8 for a list of building codes.)

The live-load values for occupancy tend to be conservative. Most building codes permit a percentage reduction when the contributing floor area is very large because it is unlikely that full live loads will be applied to the entire floor area at the same time.

As given in several building codes, the percentage reduction for live loads of 100 psf or less on members supporting 150 ft^2 or more is the *least* of the following:

$$R = 0.08A \qquad (\text{where } A > 150 \text{ ft}^2)$$

$$R = 23\left(1 + \frac{D}{L}\right)$$

$$R \leq 60$$

where R = reduction, percent

                               A = contributing area, ft^2

                               D = dead load, psf

                               L = live load, psf

No reduction is permitted in public assembly areas, garages, and roofs.

Example 2-1

The framing plan for part of a building floor is shown in Fig. 2-1. The floor
is 5-in.-thick reinforced concrete and it supports a live load of 100 psf. What
is the reduced live load for floor beam B_1 and column C_2 two floors down
from the roof?

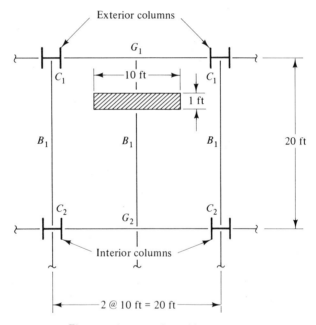

Floor continues on three sides

Figure 2-1   Example 2-1.

Solution.   Concrete weighs 150 lb/ft³. Therefore, a 5-in.-thick reinforced concrete floor has a dead load per square foot of

$$\frac{5(150)}{12} = 62.5 \text{ psf}$$

The live load is given as 100 psf.

Reduced live load for Beam B_1. The contributing area per *foot of length* for the beam is 1 ft by 10 ft, as shown by the cross-hatched area in Fig. 2-1. Thus the contributing area for the beam is 10 ft by 20 ft, or 200 ft^2. The percentage reduction is calculated as follows:

$$R = 0.08(200) = 16\% \leq 60$$

$$R = 23\left(1 + \frac{62.5}{100}\right) = 37.4\% \leq 60$$

Using the smaller value of R, the reduced live load becomes

$$L = 100(1 - 0.16) = 84\,\text{psf}$$

Reduced live load for Column C_2. The contributing area for each column is 20 ft by 20 ft, or 400 ft^2. Two floors down it will be multiplied by 2. The percentage reduction is calculated as follows:

$$R = 0.08(2)(400) = 64\% \leq 60$$

$$R = 23\left(1 + \frac{62.5}{100}\right) = 37.4\% \leq 60$$

Using the smaller value of R, the reduced live load becomes

$$L = 100(1 - 0.374) = 62.6\,\text{psf} \quad (\text{say, 63 psf})$$

2-4 SNOW LOADS

The principal live loads on roofs are snow and wind. The weight of snow to be considered varies with such factors as wind speed and direction, the location, and the slope of the roof. Along with snow the possibility of ice must also be considered. Partial melting of the snow followed by freezing and more snow or freezing rain can cause heavy roof loads.

The collapse of hundreds of roofs in northern areas of the United States during recent winters has been attributed to unprecedented amounts of snow and nonuniform loads caused by drifting snow. The collapsed roofs in most cases were designed to conform to the minimum snow loads prescribed by the governing building code. As a result of these failures, the Snow Load Subcommittee of ANSI has drafted new standards for snow loads.

The ANSI code gives a basic snow load q, corresponding with the ground load. The basic snow load is then multiplied by the appropriate snow-load coefficient C_s to account for wind speed and direction, slope of roof, and nonuniform snow accumulations. The basic snow load is shown on maps of the United States by contour lines that indicate equal ground snow loads in psf. Depending on the degree of human risk associated with roof failure, the applicable map is based on mean recurrence intervals for 25, 50, or 100 years.

The basic snow-load coefficient is increased or decreased to reflect various factors, such as nonuniform snow accumulations, snow slide-off from roofs, and the formation of ice on roofs due to the temperature difference between the inside and outside of the building.

The minimum design loads recommended by the Housing and Home Finance Agency provides an *alternative*, uncomplicated method for including the effect of snow on a roof. These recommendations, which are based on studies of the U.S. Weather Bureau, are given in Table 2-2. The values given are the recommended minimum snow-load distribution on the *horizontal projection* of the roof area.

TABLE 2-2   Minimum Design Snow Load[a]

	Roof Slope			
Region	0.25 or less	0.5	0.75	1.0 or more
Southern states	20	15	12	10
Central states	25	20	15	10
Northern states	30	25	15	10
Great Lakes, New England, and mountain areas[b]	40	30	20	10

[a]Given in psf on horizontal projection of roof.
[b]Areas include northern portions of Massachusetts, Michigan, Minnesota, Wisconsin, and New York; Maine, New Hampshire, and Vermont; the Appalachians above 2000 ft elevation, Pacific coast mountains above 1000 ft, and Rocky Mountains above 4000 ft.

Example 2-2

For the structure shown in Fig. 2-2, determine the snow load using results from the Housing and Home Finance Agency study. The structure is located in Richmond, Virginia.

Solution.   The slope of the roof is 18 to 36, or 0.5, and Virginia is a southern state. From Table 2-2 the minimum design load is 15 psf = 0.015 ksf for a horizontal projection of the roof. This is illustrated in Fig. 2-3.

The loads transmitted to end-panel points (1) of the truss by the eave struts are given by

$$0.015(4.5)22 = 1.485 \text{ k}$$

and the loads transmitted to the interior panel points (2, 3, and 4) by the purlins are given by

$$0.015(9)(22) = 2.97 \text{ k}$$

The loads for the various panel points of the truss are shown in Fig. 2-4.

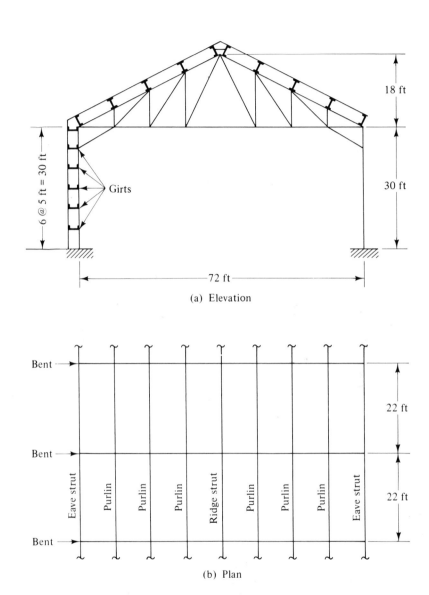

(a) Elevation

(b) Plan

Figure 2-2   Example 2-2.

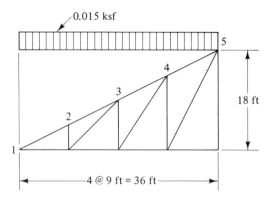

Figure 2-3 Example 2-2.

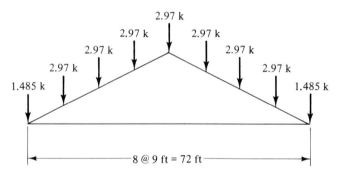

Figure 2-4 Example 2-2.

2-5 WIND LOADS

For some structures, such as tall chimneys, suspension-type structures, and high-rise buildings, wind is one of the most important loads. For such structures, the wind loads are difficult to predict and wind-tunnel studies may be required.

Wind is the free movement of air. It occurs in gusts rather than a steady flow. A gust (fluctuations in wind speed) may vary from 10 ft/sec to 100 ft/sec (7 to 70 mph) during a thunderstorm. The direction and speed varies with the height above the ground. Wind is also affected by buildings and other topographic features, such as variations in ground elevation.

The determination of the force of the wind on a structure in its path is a complex problem. The usual practice is to convert the dynamic wind pressure into an equivalent static pressure. The dynamic pressure is a function of the velocity of the wind and the density of air. That is,

$$q = 0.00256 \, V^2$$

where q = dynamic pressure, psf
      V = speed of the wind, mph
The constant includes the density of the air. Dynamic pressure is converted to static pressure by the expression

$$p = Cq$$

where p = static pressure, psf
      q = dynamic pressure, psf
The coefficient C depends on such factors as *shape and size* of the structure, wind *gusts*, and the *shielding effect* from adjacent structures or ground cover.

The minimum design loads recommended by the *National Building Code* for horizontal wind on the vertical surfaces of a small rectangular-shaped structure with heights of 100 ft or less are given in Table 2-3. The loads given include the effect of shape and gusts. No allowance was made for the shielding effect.

For roofs with slopes of less than 30°, the roof pressure is a suction (outward—perpendicular to the roof surface) equal to 1.25 times the pressure given in Table 2-3. For slopes greater than 30°, the roof pressure is given in the table.

TABLE 2-3   **National Building Code (National) Wind Requirements, Pressure**[a]

Height Zone (ft)	Wind Velocity		
	Minimum	Moderate	Severe[b]
Less than 30	15	25	35
30–49	20	30	45
50–99	25	40	55
100–499	30	45	70

[a]Horizontal wind on vertical surface, psf.
[b]Use within 50 miles of the Gulf coast, and the Atlantic coast from Chesapeake Bay to Key West, Florida.

Example 2-3

For the structure shown in Example 2-2 (Fig. 2-2), determine the wind loads using the NBC code. Assume that Richmond is a moderate wind-storm area.

Solution.   The average height h for the roof as shown in Fig. 2-2 is given by

$$h = 30 + \frac{18}{2} = 39 \text{ ft}$$

The wind pressure p from Table 2-3 for a moderate wind-storm area is

$$p = 30 \text{ psf} = 0.030 \text{ ksf}$$

The slope of the roof is 18 to 36, or 26.6°, which is less than 30°. Therefore, the pressure on the roof is

$$p' = -1.25(0.030) = -0.0375 \text{ ksf}$$

For a column height of 30 ft the pressure p for the columns is given by

$$p = 25 \text{ psf} = 0.025 \text{ ksf}$$

The pressures are illustrated in Fig. 2-5.

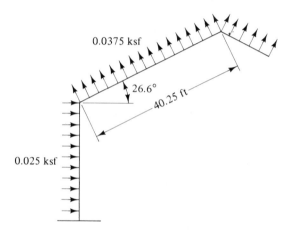

Figure 2-5   Example 2-3.

The loads transmitted to end panel points of the truss by the eave struts are given by

$$\frac{0.0375(40.25)(22)}{8} = 4.15 \text{ k}$$

and the loads transmitted to the interior panel points by the purlins are twice that value, or 8.30 k. The loads transmitted to the top of the columns by the girts are given by

$$\frac{0.025(30)(22)}{12} = 1.375 \text{ k}$$

and the loads transmitted to the columns by the girts are twice that value, or 2.75 k.

These loads are shown in Fig. 2-6.

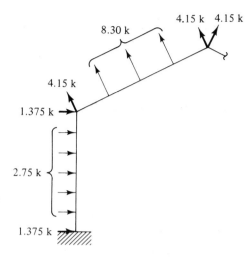

8.30 k

4.15 k 4.15 k

4.15 k

1.375 k →

2.75 k

1.375 k

Figure 2-6   Example 2-3.

2-6 IMPACT

Impact is produced by dynamic loading. The effect of dynamic loading on a structure depends not only on the magnitude of the load but also on the mass of the structure and the manner in which the load is applied. Examples of impact loading in buildings are found in elevators, traveling cranes, shaft- or motor-driven light machinery, reciprocating machinery, and floor and balcony hangers.

To account for the dynamic or impact effects, many specifications call for a percentage increase in the live load. The percentage increase is called the *impact factor*. AISC Specification 1.3.3 provides that if not otherwise specified, the increase in the live load should be:

For support of elevators .... 100 percent
For cab-operated traveling support
    girders and their connections .... 25 percent
For pendant-operated traveling crane
    support girders and their connections .... 10 percent
For supports of light machinery, shaft
    or motor driven, not less than .... 20 percent
For supports of reciprocating machinery
    or power driven units, not less than .... 50 percent
For hangers supporting floors and
    balconies .... 33 percent

Example 2-4

Determine the design load for a hanger supporting a balcony if the dead load is 15 k and the live load is 40 k. Consider impact.

Solution. From AISC Specification 1.3.3 for hangers supporting floors or balconies, the impact factor is 33 percent. The live load should be increased by 33 percent.

$$\text{Dead load} \qquad\qquad = 15.0\,\text{k}$$
$$\text{Live load} = 1.33(40) = \underline{53.2\,\text{k}}$$
$$\text{Design load} \qquad\qquad\; 68.2\,\text{k}$$

2-7 EARTHQUAKE FORCES

In areas of the world where earthquakes may occur, their effects must be considered in the design of structures. An earthquake causes ground motion in all directions. The ground motion may be divided into horizontal and vertical components. The vertical component of the ground motion is small and can usually be neglected. The horizontal component is much larger and produces the earthquake load. When the ground under a structure having a mass moves suddenly, the inertia of the mass (inertial reaction) tends to resist this ground motion (Fig. 2-7). A horizontal shear force V is developed between the ground and the structure. In many codes empirically prescribed values of the shear force are assumed to be equal to the inertial reaction. This approach may be adequate for simple low-rise buildings, but for more complex high-rise buildings a complete dynamic analysis is required.

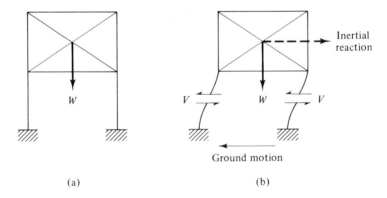

Figure 2-7 (a) Ground at rest; (b) ground in horizontal motion due to earthquake.

2-8 DESIGN LOAD COMBINATIONS

The loads supported by roof trusses and one-story building frames include dead loads, snow loads, and wind loads. Full wind would probably cause a partial snow load due to drifting. Load combinations used in the design of *roofs* may include the following:

1. Dead load and snow load
2. Dead load and wind load
3. Dead load, wind load, and a partial snow load

Ice might also be considered in each of the combinations above, including wind load.

Floor loads for buildings consist of dead loads and live loads. When the contributing floor is very large, some specifications permit a percentage reduction in the live load (see Sec. 2-3). Therefore, *floor design* often includes the following load combinations:

1. Dead load and live load
2. Dead load and partial live load

For multistory building frames, dead loads, live loads, wind loads, and where applicable, earthquake loads must be considered. Combinations must be selected that produce maximum design loads. For example, live loads over alternate bays of each floor of a building may produce maximum load conditions for the floor beams. Thus *multistory building design* may include the following load combinations:

1. Dead load and live load
2. Dead load and partial live load
3. Dead load, live load, and wind load
4. Dead load, live load, and earthquake load

Engineering judgment is required in the selection of the proper load combinations.

AISC Specification 1.5.6 permits an increase of one-third in the allowable stresses when wind or earthquake forces are included in the load combination. This specification applies to load combinations 2 and 3 for roofs, and load combinations 3 and 4 for multistory buildings.

PROBLEMS

Use the *National Building Code* for wind loads and recommended snow loads from the Housing and Home Finance Agency.

2-1. The framing plan for part of a building floor is shown in Fig. 2-1. The floor is a 4-in.-thick reinforced concrete slab and it supports a live load of 80 psf.

Determine the reduced live load for the following:
(a) Floor beam B_1.
(b) Column C_2 two stories down from the top.

2-2. Part of the floor-framing plan is shown in the figure. The floor is a 4-in.-thick concrete slab and it supports a live load of 80 psf. Determine the reduced live load for the following:
(a) Floor beams B_2, B_3, and B_4.
(b) Exterior column C_2 and interior column C_4 two stories down from the top.

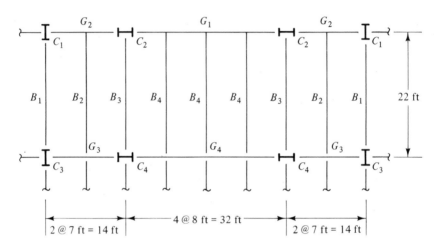

Prob. 2-2

2-3. Repeat Prob. 2-2 with a 6-in.-thick concrete slab and a live load of 100 psf.

2-4. The framing plan for beams and girders used to support a reinforced concrete floor is shown in the figure. What is the reduced live load for beams B_1 and column C_1 two stories down from the roof if the live load, floor thickness, and spans L_1 and L_2 are as follows?
(a) $L = 80$ psf, $t = 4$ in., $L_1 = 18$ ft, $L_2 = 20$ ft.
(b) $L = 80$ psf, $t = 5$ in., $L_1 = 20$ ft, $L_2 = 22$ ft.
(c) $L = 100$ psf, $t = 6$ in., $L_1 = 22$ ft, $L_2 = 26$ ft.
(d) $L = 100$ psf, $t = 6$ in., $L_1 = 26$ ft, $L_2 = 28$ ft.

2-5. For the structure shown in Example 2-2 (Fig. 2-2), determine the snow loading if it is located in northern Minnesota.

2-6. What are the snow loads from the purlins, eave struts, and ridge struts (points 1 through 9) on the building frame shown in the figure? The structure is located in Atlanta, Georgia. Interpolate to find design snow loads for roof slopes between tabulated values.

2-7. What are the snow loads from the purlins and eave struts (points 1 through 7) on the building frame shown in the figure? The structure is located in Kansas City, Missouri.

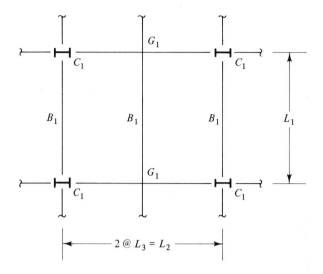

Prob. 2-4

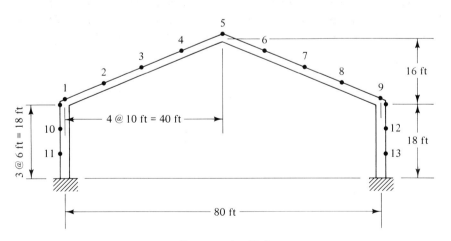

Frame spacing 20 ft

Prob. 2-6

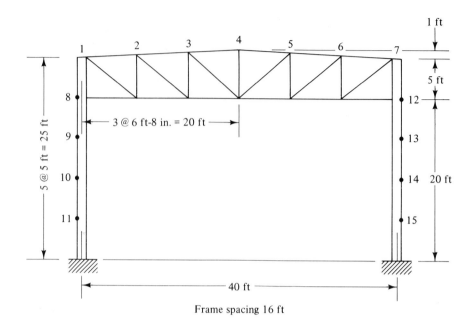

Frame spacing 16 ft

Prob. 2-7

2–8. For the structure shown in Example 2-2 (Fig. 2-2), determine the wind loads if the structure is in New Orleans, Lousiana.

2-9. For the building frame shown in Prob. 2-6, determine the wind loads for the purlins, etc. (points 1 through 13). Assume a moderate wind-storm area.

2-10. For the building frame shown in Prob. 2-7, determine the wind loads for the purlins, etc. (points 1 through 15). Assume a severe wind-storm area.

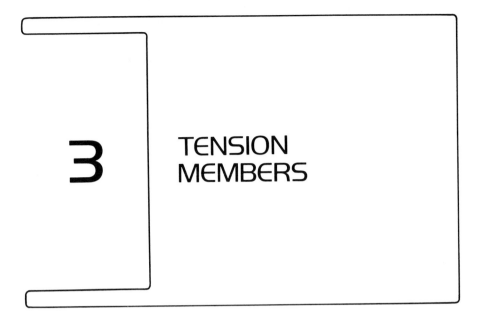

3 TENSION MEMBERS

3-1 INTRODUCTION

A tension member is one that is acted on by an axial tensile force only. Tension members are found in most steel structures as main load-carrying members or as secondary bracing members. They may consist of a single structural shape or be built up from several structural shapes.

The selection of a tension member is one of the simplest problems in steel design. The member remains straight under a tensile load and the stress produced is a simple tensile stress. The cross-sectional area of the member must be proportioned so that the load does not produce a stress greater than the allowable tensile stress. The member must also have sufficient stiffness so that it does not sag or vibrate.

3-2 TYPES OF TENSION MEMBERS

Some of the more common types of tension members are rods, bars, cables, and such rolled shapes as angles, channels, tees, and W or S shapes. They may also be built up of rolled shapes and plates. The selection of the type of tension member very often depends on the kind of end connection used.

Cross sections for a few of the many possible tension members are shown in Fig. 3-1. The round rod, flat bar, and angle shown in (a), (b), and (c) are commonly used for hangers, tie rods, and bracing. Building trusses can be fabricated from the double angles, double channels, W or S shapes, and tees

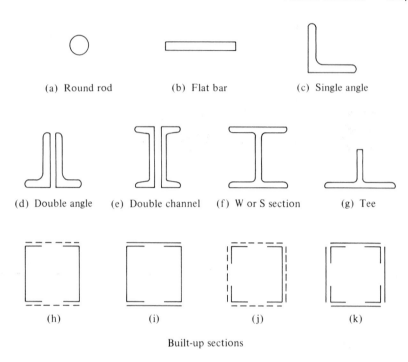

(a) Round rod   (b) Flat bar   (c) Single angle

(d) Double angle   (e) Double channel   (f) W or S section   (g) Tee

(h)   (i)   (j)   (k)

Built-up sections

Figure 3-1 Typical sections used as tension members.

shown in (d), (e), (f), and (g). The built-up sections in (h), (i), (j), and (k) are suitable for bridges.

3-3 ALLOWABLE TENSILE STRESS

The tensile stress produced by a tensile axial load on a straight uniform member is assumed to be uniform over the cross-sectional area. The tensile stress is given by

$$f_t = \frac{P}{A} \tag{3-1}$$

where f_t = tensile stress, ksi
  P = axial load, k
  A = cross-sectional area, in.^{2}

The axial load acts along the straight uniform member at the centroid of the cross-sectional area.

The entire cross-sectional area of the member is called the *gross area* A_g. A cross-sectional area where holes are present must be reduced by the area of the holes and is called the *net area* A_n.

Tests to failure on tension members show that failure may occur either by yielding of the gross section or fracture of the net section. Thus the AISC *Specifications* provide allowable stresses for the gross area based on the yield stress and for the net area based on the ultimate or tensile stress. These are given in AISC Specification 1.5.1.1 as follows: Except for pin-connected members, the allowable tensile stress on the *gross area*

$$F_t = 0.60 F_y \qquad (3\text{-}2)$$

and on the *effective net area*

$$F_t = 0.50 F_u \qquad (3\text{-}3)$$

where F_y = yield stress, ksi
$\quad\ F_u$ = minimum ultimate tensile stress, ksi

3-4 NET AREA

The net area is the gross cross-sectional area of the member from which the area of the holes has been subtracted. The area to be subtracted for the holes depends on the method used to make the holes. Three methods are used: punching, subpunching and reaming, and drilling. In the usual method standard holes are punched $\frac{1}{16}$ in. larger than the diameter of the hole (AISC Specification 1.23.4.1). Metal on the edge of the hole is damaged by the punching operation. This is accounted for in design by adding $\frac{1}{16}$ in. to the diameter of the hole (AISC Specification 1.14.4). Thus the width to be deducted for standard holes is the bolt or rivet diameter plus $\frac{1}{8}$ in. In the second method the holes are punched undersized and reamed to the finished size after the members are joined together. In the third method the holes are drilled $\frac{1}{32}$ in. larger than the bolt or rivet.

Larger-than-standard-sized holes can be used with high-strength bolts larger than $\frac{5}{8}$ in. The holes may also be oversize, short slotted, and long slotted to facilitate erection.

Example 3-1   *Minimum Net Area*

What is the net area A_n for the tension member shown in Fig. 3-2? Standard holes for $\frac{3}{4}$-in. bolts are used.

Solution.   The gross area and net area are calculated as follows:

$$A_g = 9\left(\frac{5}{8}\right) = 5.625 \text{ in.}^2$$

The width to be deducted for holes is $3(\frac{3}{4} + \frac{1}{8}) = 2.625$ in. Therefore, the net area

$$A_n = A_g - (\text{width of holes})\,(\text{thickness of plate})$$

$$= 5.625 - 2.625\left(\frac{5}{8}\right) = 3.98 \text{ in.}^2$$

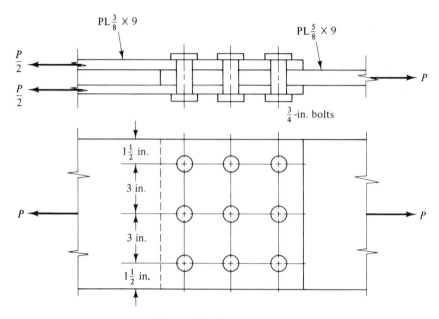

Figure 3-2 Example 3-1.

3-5 EFFECT OF STAGGERED HOLES

When there are more than one line or bolt or rivet holes in a member, the holes may be staggered, that is, not lined up in a direction transverse to the direction of the load. In such a case, more than one possible failure section may occur. The section with the minimum net area is the failure section.

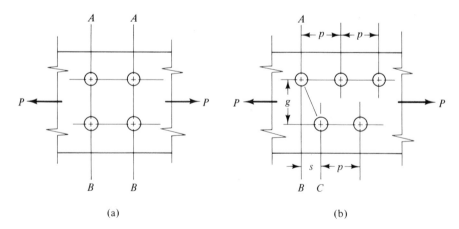

(a)   (b)

Figure 3-3 Paths of failure on net section.

Consider the tension members shown in Fig. 3-3. The critical or failure section in (a) is along section $A - B$. In (b) the failure section may be through one hole (section $A - B$) or through two holes along the staggered section $A - C$. AISC Specification 1.14.2 accounts for the difference between paths $A - B$ and $A - C$ in (b) by a length correction

$$\frac{s^2}{4g}$$

where s = stagger or spacing of any two successive holes parallel to the loading direction, in.

$\qquad g$ = gage distance transverse to the loading direction of the same two holes, in.

The net area for sections $A - B$ and $A - C$ are as follows:

$$A - B: A_n = \left[\text{width } A - B - \left(\text{bolt diam.} + \frac{1}{8}\text{in.} \right) \right]$$

$$\times \text{(plate thickness)}$$

$$A - C: A_n = \left[\text{width } A - B - 2\left(\text{bolt diam.} + \frac{1}{8}\text{in.} \right) + \frac{s^2}{4g} \right]$$

$$\times \text{(plate thickness)}$$

The section with the minimum net area is the failure section.

Example 3-2   *Minimum Net Area: Stagger*

Determine the net area of the $\frac{3}{4} \times 14$ PL shown in Fig. 3-4. The holes are for $\frac{7}{8}$-in. bolts.

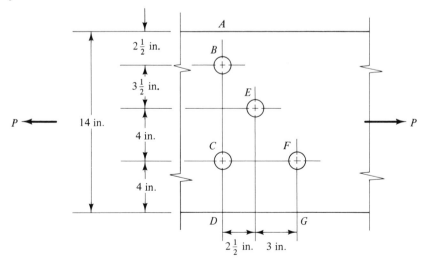

Figure 3-4   Example 3-2.

Solution. Various cross sections are calculated.

$$ABCD: \left[14 - 2\left(\frac{7}{8} + \frac{1}{8}\right)\right]\frac{3}{4} = 9.00 \text{ in.}^2$$

$$ABECD: \left[14 - 3\left(\frac{7}{8} + \frac{1}{8}\right) + \frac{(2.5)^2}{4(3.5)} + \frac{(2.5)^2}{4(4)}\right]\frac{3}{4} = 8.88 \text{ in.}^2$$

$$ABEFG: \left[14 - 3\left(\frac{7}{8} + \frac{1}{8}\right) + \frac{(2.5)^2}{4(3.5)} + \frac{(3)^2}{4(4)}\right]\frac{3}{4} = 9.01 \text{ in.}^2$$

The minimum net area or control is 8.88 in.^{2}.

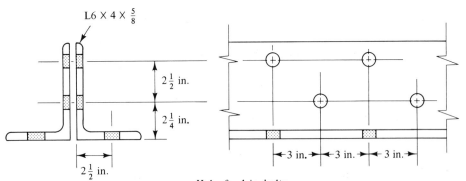

L6 × 4 × $\frac{5}{8}$

$2\frac{1}{2}$ in.

$2\frac{1}{4}$ in.

$2\frac{1}{2}$ in.

3 in.   3 in.   3 in.

Holes for 1-in. bolts

(a)

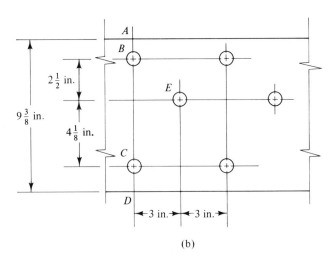

$9\frac{3}{8}$ in.

$2\frac{1}{2}$ in.

$4\frac{1}{8}$ in.

A
B
E
C
D

3 in.   3 in.

(b)

Figure 3-5   Example 3-3.

Example 3-3   *Minimum Net Area: Stagger*

Determine the minimum net area for the two L6 × 4 × $\frac{5}{8}$ shown in Fig. 3-5(a). The holes are for 1-in. bolts.

Solution.   For calculation purposes one angle can be opened into a plate with the length along the centerline of the angle maintained as shown in Fig. 3-5(b). The gross width of the plate is the sum of the leg lengths less the thickness ($6 + 4 - \frac{5}{8} = 9\frac{3}{8}$ in.), and the gage for holes in opposite legs is the sum of the gages from the back of the angle less the thickness ($2\frac{1}{4} + 2\frac{1}{2} - \frac{5}{8} = 4\frac{1}{8}$ in.). Various cross sections of the plate are calculated as follows:

$$ABCD: \left[9.375 - 2\left(1 + \frac{1}{8}\right)\right]\frac{5}{8} = 4.45 \text{ in.}^2$$

$$ABECD: \left[9.375 - 3\left(1 + \frac{1}{8}\right) + \frac{(3)^2}{4(2.5)} + \frac{(3)^2}{4(4.125)}\right]\frac{5}{8} = 4.65 \text{ in.}^2$$

The minimum net area or control for one angle is 4.45 in.^{2}. For two angles the minimum net area or control is 8.90 in.^{2}.

3-6 EFFECTIVE NET AREA

A bolted or riveted connection that is attached to some but not all of the cross-sectional elements of a tension member causes a nonuniform stress distribution over the net cross-sectional area of the member. Connections made to one leg of an angle section or the flanges of W, S, or C shapes are examples of this type of connection.

The nonuniform transfer of stress at a connection is discussed in AISC Commentary 1.14.2.2. The cause of the nonuniformity is attributed to a condition often referred to as *shear lag*.

To account for this nonuniform stress distribution, AISC Specification 1.14.2.2 provides an *effective net area*. The effective net area is calculated from the formula

$$A_e = C_t A_n$$

where A_e = net effective area, in.^{2}

C_t = reduction coefficient

A_n = net area, in.^{2}

Values of the reduction coefficient C_t are given in Table 3-1.

TABLE 3-1   Effective Net Area for Various Types of Tension Members

Type of Member	Special Requirements	Effective Net Area, A_e
(a) Load transmitted by bolts or rivets through all cross-sectional elements of the member	None	A_n
(b) Short fittings such as splice plates, gusset plates, or beam-to-column fittings	None	A_n but $A_n \leq 0.85 A_g$*
(c) W, M, or S shapes with flange width $\geq \frac{2}{3}$ section depth and structural tees cut from these shapes	1. Connection to the flange or flanges 2. No fewer than three fasteners per line in the direction of stress	$0.90 A_n$
(d) W, M, or S shapes with flange width $< \frac{2}{3}$ section depth and structural tees cut from these shapes; all other shapes, including built-up cross sections	No fewer than three fasteners per line in the direction of stress	$0.85 A_n$
(e) All members in (c) or (d)	Two fasteners per line in the direction of stress	$0.75 A_n$

*AISC Specification 1.14.2.3.
Source: Adapted from AISC Specification 1.14.2.2.

3-7 LIMITING SLENDERNESS FOR TENSION MEMBERS

A tension member must be designed with sufficient strength to support the axial load. In addition, the member must also have sufficient stiffness or rigidity to prevent excessive vibration of the member, strengthen against small lateral loads, and minimize sagging due to the weight of the member.

Stiffness or rigidity is measured by the slenderness ratio. The slenderness ratio is equal to L/r, where L is the length of the member between lateral supports and r is the *least* radius of gyration of the cross section of the member.

A slender flexible member has a high L/r ratio and a stiff member has a low L/r ratio.

AISC Specification 1.8.4 recommends that except for rods, the slenderness ratio preferably should not exceed 240 for *main tension members* and 300 for lateral *bracing members* and other *secondary members*.

From our study of strength of materials we recall that the radius of gyration r of a section with respect to an axis is given by $r = \sqrt{I/A}$, where I is the moment of inertia of the section with respect to that axis and A is the cross-sectional area of the section. The radius of gyration is tabulated for standard structural shapes in the AISC *Manual*. For built-up members it must be calculated.

Example 3-4   *Slenderness Ratios: Various Structural Shapes*

Determine the slenderness ratio for the tension members with cross sections and lengths as shown in Fig. 3-6.

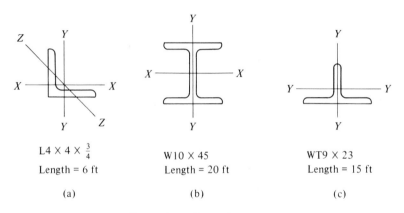

L4 × 4 × $\frac{3}{4}$
Length = 6 ft

(a)

W10 × 45
Length = 20 ft

(b)

WT9 × 23
Length = 15 ft

(c)

Figure 3-6   Example 3-4.

Solution.   Values for the radius of gyration are taken from the AISC *Manual*.
(a) L4 × 4 × $\frac{3}{4}$: $r_x = r_y = 1.19$ in., $r_z = 0.778$ in.

$$\frac{L}{r_{min}} = \frac{L}{r_z} = \frac{6(12)}{0.778} = 92.5$$

(b) W10 × 45: $r_x = 4.32$ in., $r_y = 2.01$ in.

$$\frac{L}{r_{min}} = \frac{L}{r_y} = \frac{20(12)}{2.01} = 119.4$$

(c) WT9 × 23: $r_x = 2.77$ in., $r_y = 1.29$ in.

$$\frac{L}{r_{min}} = \frac{L}{r_y} = \frac{15(12)}{1.29} = 139.5$$

Example 3-5   *Slenderness Ratios: Built-up Shapes*

Determine the slenderness ratio for a tension member with cross section and length as shown in Fig. 3-7.

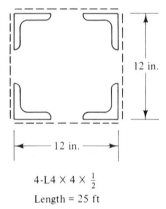

4-L4 × 4 × $\frac{1}{2}$

Length = 25 ft   **Figure 3-7**   Example 3-5.

Solution.   The values for the moments of inertia and location of the centroid for one L4 × 4 × $\frac{1}{2}$ are from the AISC *Manual.*

$$L4 \times 4 \times \tfrac{1}{2}: I_x = I_y = 5.56 \text{ in.}^4, \, x = y = 1.18 \text{ in.}$$

For the complete cross section of the member,

$$I_x = I_y = 4[5.56 + 3.75(6 - 1.18)^2] = 370.7 \text{ in.}^4$$

$$A = 4(3.75) = 15 \text{ in.}^2$$

$$r_x = r_y = \sqrt{\frac{370.7}{15}} = 4.97 \text{ in.}$$

$$\frac{L}{r_{\min}} = \frac{25(12)}{4.97} = 60.3$$

3-8 DESIGN OF TENSION MEMBERS

The type of member selected in design is usually dependent on the connections used for the structure. If the structure is welded, no holes need to be deducted from the gross area except for possible holes used during erection. Of course, these holes should be considered in the design.

The required cross-sectional area for a tension member must be determined so that the tensile stress does not exceed the allowable values. To find this area, we rewrite Eq. (3-1):

$$\text{required } A \geq \frac{P}{F_t} \tag{3-4}$$

where P = axial load, k
$\quad$ F_t = allowable tensile stress, ksi

Required gross area.   The required gross area is found by substituting the allowable tensile stress $F_t = 0.6F_y$ in Eq. (3-4). The

$$\text{required } A_g \geq \frac{P}{0.6F_y} \qquad (3\text{-}5)$$

Required effective net area.   Substituting $F_t = 0.5F_u$ in Eq. (3-4), we have

$$\text{required } A_e \geq \frac{P}{0.5F_u} \qquad (3\text{-}6)$$

Required net area.   Since $A_e = C_tA_n$, the required net area from Eq. (3-6) is given by

$$\text{required } A_n \geq \frac{P}{0.5F_uC_t} \qquad (3\text{-}7)$$

Required gross area.   A second equation for the required gross area is found by adding the hole area and subtracting the area due to stagger from the required net area given by Eq. (3-7):

$$\text{required } A_g \geq \frac{P}{0.5F_uC_t} + A_{\text{holes}} - A_{\text{stagger}} \qquad (3\text{-}8)$$

where the areas due to the holes and stagger are estimated.
$\quad$ A tension member must have a gross area that satisfies Eqs. (3-5) and (3-8). In addition, the member should have adequate stiffness.

Required minimum radius of gyration.   The required minimum radius of gyration for a main member can be found from the recommendation of AISC Specification 1.8.4 that the slenderness ratio should not exceed 240. Thus

$$\text{required } r_{\text{min}} \geq \frac{L}{240} \qquad (3\text{-}9)$$

where L is the length of the member, in.

$\quad$ The following examples illustrate the design of tension members with various cross sections acted on by dead loads, live loads, impact, and wind loads.

Example 3-6   *Single- and Double-Angle Tension Member*

A tension member has a length of 15 ft and is stressed by a dead load of 44 k and a live load of 59 k. Assume that the member is a main member and $\frac{7}{8}$-in.-diameter bolts are used with at least three bolts in each row.

(a) Select the lightest single-angle tension member. Use two rows of bolts.
(b) Select the lightest double-angle tension member. Use one row of bolts with the long legs back to back separated by a $\frac{3}{8}$-in. space.

Solution. Design loads:

$$\text{Dead load} = \quad 44\,\text{k}$$

$$\text{Live load} = \quad \underline{59\,\text{k}}$$

$$103\,\text{k}$$

From Table 1-1 for A36 steel: $F_y = 36$ ksi, $F_u = 58$ ksi. From Eqs. (3-5) and (3-6) we have

$$\text{required } A_g \geq \frac{P}{0.6F_y} = \frac{103}{22} = 4.68 \text{ in.}^2$$

$$\text{required } A_e \geq \frac{P}{0.5F_u} = \frac{103}{0.5(58)} = 3.55 \text{ in.}^2$$

The effective net area $A_e = 0.85A_n$ for type of member (d) of Table 3-1 (AISC Specification 1.14.2.2). Therefore, the

$$\text{required } A_n \geq \frac{A_e}{C_t} = \frac{3.55}{0.85} = 4.178 \text{ in.}^2$$

To satisfy Eq. (3-9) (AISC Specification 1.8.4), the

$$\text{required } r_{\min} \geq \frac{L}{240} = \frac{15(12)}{240} = 0.75 \text{ in.}$$

(a) Selection of a single-angle member. With two rows of bolts the required gross area from Eq. (3-8) will depend on the area of two standard holes. The cross-sectional area of two holes $A_h = n(d + \frac{1}{8})t = 2(\frac{7}{8} + \frac{1}{8})t$, where t is the thickness of the angle selected. In Table (a) various listed angle thicknesses were investigated to find the lightest angle with equal legs and unequal legs that provide sufficient gross area. The lightest angles were selected from Table (a) for the design. Use L6 × 6 × $\frac{7}{16}$ with equal legs or L7 × 4 × $\frac{1}{2}$ with unequal legs.

(b) Selection of a double-angle member. With two angles and one row of bolts, the gross area from Eq. (3-8) will depend on the area of two standard holes. The cross-sectional area of two holes $A_h = n(d + \frac{1}{8})t = 2(\frac{7}{8} + \frac{1}{8})t$, where t is the thickness of the angle selected. The angle selections are shown in Table (b). Use two L4 × 4 × $\frac{5}{16}$ with equal legs or two L5 × 3 × $\frac{5}{16}$ with unequal legs.

TABLE (a) for Example 3-6   Angle Selection

Angle Thickness, t	Area of Holes, A_h	Required A_g Larger of Eqs. (3-5) and (3-8)	Selection of Lightest Equal and Unequal Leg Angles and Properties
$\frac{3}{8}$	0.75	4.93	Not available
$\frac{7}{16}$	0.875	5.05	[a,b]L6 × 6 × $\frac{7}{16}$, A = 5.03 in.^{2}, r_z = 1.19 in. [b]L8 × 6 × $\frac{7}{16}$, A = 5.93 in.^{2}, r_z = 1.31 in.
$\frac{1}{2}$	1.0	5.18	L6 × 6 × $\frac{1}{2}$, A = 5.75 in.^{2}, r_z = 1.18 in. [a]L7 × 4 × $\frac{1}{2}$, A = 5.25 in.^{2}, r_z = 0.872 in.
$\frac{5}{8}$	1.25	5.43	[b]L6 × 4 × $\frac{5}{8}$, A = 5.86 in.^{2}, r_z = 0.867 in. [b]L5 × 5 × $\frac{5}{8}$, A = 5.86 in.^{2}, r_z = 0.978 in.
$\frac{3}{4}$	1.5	5.68	L5 × 5 × $\frac{3}{4}$, A = 6.94 in.^{2}, r_z = 0.975 in. L5 × $3\frac{1}{2}$ × $\frac{3}{4}$, A = 5.81 in.^{2}, r_z = 0.748 in.[c]

[a]Design selection.
[b]Angle may not be readily available.
[c]$(r_z)_{req} \geq 0.75$ in.

Table (b) for Example 3-6   Angle Selection

Angle Thickness, t	Area of Holes, A_h	Required A_g Larger of Eqs. (3-5) and (3-8)	Selection of Lightest Equal and Unequal Leg Angles and Properties
$\frac{1}{4}$	0.5	4.68	Not available
$\frac{5}{16}$	0.625	4.80	[a]2 − L5 × 3 × $\frac{5}{16}$, A = 4.80 in.^{2}, r_y = 1.22 in. [a]2 − L4 × 4 × $\frac{5}{16}$, A × 4.80 in.^{2}, r_x = 1.24 in.
$\frac{3}{8}$	0.75	4.93	2 − L4 × 3 × $\frac{3}{8}$, A = 4.97 in.^{2}, r_x = 1.26 in. 2 − L$3\frac{1}{2}$ × $3\frac{1}{2}$ × $\frac{3}{8}$, A = 4.97 in.^{2}, r_x = 1.07 in.
$\frac{1}{2}$	1.0	5.18	2 − L4 × 3 × $\frac{1}{2}$, A = 6.50 in.^{2}, r_x = 1.25 in. 2 − L3 × 3 × $\frac{1}{2}$, A = 5.50 in.^{2}, r_x = 0.898 in.

[a]Design selection.

Example 3-7   W Section Tension Member: Impact

Design a W10 section tension member 20 ft long to support a dead load of 115 k and live load of 75 k using A572 Grade 50 steel. A connection is made to the flanges by two lines of bolts in each flange with at least three $\frac{7}{8}$-in.-diameter bolts in each line. (There are four bolt holes at any cross section.) The tension member is a hanger supporting a floor. Consider impact.

Solution. The live load should be increased by 33 percent for impact on hangers supporting floors and balconies (AISC Specification 1.3.3).

$$\begin{array}{ll} \text{Dead load} & = 115 \text{ k} \\ \text{Live load} = 1.33(75) = & \underline{100 \text{ k}} \\ & 215 \text{ k} \end{array}$$

From Table 1-1 for A572 Grade 50 steel: $F_y = 50$ ksi, $F_u = 65$ ksi. From Eqs. (3-5) and (3-6) the

$$\text{required } A_g \geq \frac{P}{0.6F_y} = \frac{215}{0.6(50)} = 7.17 \text{ in.}^2$$

$$\text{required } A_e \geq \frac{P}{0.5F_u} = \frac{215}{0.5(65)} = 6.615 \text{ in.}^2$$

Assume that the flange width of the W10 is greater than two-thirds of the section depth. The effective area $A_e = 0.90A_n$ for type of member (c) of Table 3-1 (AISC Specification 1.14.2.2). Therefore, the

$$\text{required } A_n = \frac{A_e}{0.90} = \frac{6.615}{0.90} = 7.35 \text{ in.}^2$$

To satisfy Eq. (3-9) (AISC Specification 1.8.4), the

$$\text{required } r_{\min} \geq \frac{L}{240} = \frac{20(12)}{240} = 1.0$$

Try W10 × 33: $A = 9.71$ in.^{2}, $t_f = 0.435$ in., $b_f = 7.960$ in., $d = 9.73$ in., $r_{\min} = r_y = 1.98$ in. From Table 3-1 (AISC Specification 1.14.2.2), $b_f = 7.960$ in. $> \frac{2}{3}d = \frac{2}{3}(9.73) = 6.49$ in. Therefore, $C_t = 0.90$, as assumed. The

$$\text{required } A_g \geq \text{required } A_n + n\left(d + \frac{1}{8}\right)t_f$$

$$\geq 7.35 + 4\left(\frac{7}{8} + \frac{1}{8}\right)0.435 = 9.09 \text{ in.}^2$$

Use a W10 × 33. The area and radius of gyration are both larger than required.

Example 3-8   Structural Tee Tension Member: Wind Load

Select a structural tee tension member to support a dead load of 50 k, a live load of 70 k, and a possible wind load of 50 k. The length of the member will be 25 ft. Use A36 steel. The connection is made by welds to the flanges.

Solution.

$$
\begin{aligned}
\text{Dead load} &= 50\,\text{k} \\
\text{Live load} &= \underline{70\,\text{k}} \\
&\;120\,\text{k}
\end{aligned}
$$

$$\text{Wind load} = 50\,\text{k}$$

The ratio of the wind load to the sum of the dead load and live load expressed as a percent is given by

$$\frac{50}{120}(100) = 41.7\% > 33.3\%$$

Therefore, from AISC Specification 1.5.6 the wind load governs the design of the tension member. The design load $P = 120 + 50 = 170$ k and the allowable stress F_t is increased by one-third. For A36 steel the allowable stress $F_t = 1.333(0.6F_y) = 1.333(22)$ for the load combination, including wind. From Eqs. (3-5) and (3-9) we have

$$\text{required } A_g \geq \frac{P}{F_t} = \frac{170}{1.333(22)} = 5.80 \text{ in.}^2$$

$$\text{required } r_{\min} \geq \frac{L}{240} = \frac{25(12)}{240} = 1.25 \text{ in.}$$

From a survey of the AISC *Manual* for the lightest WT section we could use a WT6 × 20, WT8 × 20, or WT9 × 20.

3-9 DESIGN FOR REPEATED LOADS: FATIGUE

Structural members are subjected to varying stresses due to repeated loading and unloading of the structure. Under such conditions the member may fail at stresses much lower in magnitude than the ultimate stress under static loads. These failures are due to fatigue.

Fatigue fracture can be described as a failure that starts as a microscopic crack which grows very slowly at first, then extends rapidly across the member to cause fracture. The cracks usually occur on the surface of the member at an imperfection or stress concentration where the repeated load produces mainly tension. The stress concentration may be due to poor welds, holes, or rough or damaged edges produced by the fabrication or erection of the structure.

Fatigue is usually not a problem in conventional steel building design. The number of repetitions of stress fall far short of the number required to produce fatigue. However, fatigue loading conditions may occur in supports for machinery and equipment and crane runways.

The AISC *Specifications* provide a simple design method for considering repeated loads. The method is based on the *number of loading cycles*, the *stress category* or type and nature of the connection, and the *stress range* or estimated difference between the maximum and minimum stresses in any loading cycle. The stress range f_{sr} is given by the equation

$$f_{sr} = \frac{P_{max} - P_{min}}{A} \tag{3-10}$$

where P_{max} = maximum load during any loading cycle, k
$\quad\;\, P_{min}$ = minimum load during any loading cycle, k
$\quad\;\;\, A$ = cross-sectional area, in.^{2}

Fatigue need not be considered if there are fewer than 20,000 loading cycles. For 20,000 or more loading cycles an allowable stress range is found from the AISC *Specifications*, Appendix B, as follows:

1. The loading condition is found from Table B1, where loading conditions 1, 2, 3, and 4 correspond to the loading cycles lower limits of 20,000, 1,000,000, 500,000, and 2,000,000.
2. The stress categories A, B, C, D, E, and F are identified from the illustrated examples of Fig. B1 and Table B2.
3. After identifying the loading conditions and the stress category, the allowable stress range F_{sr} is determined from Table B3.

Example 3-9   *Welded Structural Tee Tension Member: Fatigue*

Select a structural tee to support a design dead load of 45 k and a live load that ranges from 65 k tension to 10 k compression with 300,000 loading cycles. The length of the member is 12 ft. Use A36 steel. The fillet welded end connection is similar to case 17 in Fig. B1 of Appendix B of the AISC *Manual.*

Solution.   From Table 1-1 for steel used in structural steel shapes and plates for A36 steel: $F_y = 36$ ksi, $F_u = 58$ ksi.

   Loading range

$$
\begin{array}{ll}
\text{Dead load} & = \quad 45\text{ k} \\
\text{Tensile live load} & = \underline{\quad 65\text{ k}} \\
P_{max} & = \; 110\text{ k} \\
\text{Dead load} & = \quad 45\text{ k} \\
\text{Compressive live load} & = \underline{-10\text{ k}} \\
P_{min} & = \quad 35\text{ k}
\end{array}
$$

   Allowable stress range. The following values are from the AISC *Specifications*, Appendix B.

1. Loading condition 2 for 300,000 loading cycles from Table B1
2. Stress category E from Table B2
3. The allowable stress range $F_{sr} = 12.5$ ksi from Table B3

Required properties for member selection. From Eqs. (3-5) and (3-10), we have

$$\text{required } A_g \geq \frac{P_{max}}{0.6F_y} = \frac{110}{22} = 5.0 \text{ in.}^2$$

$$\text{required } A_g \geq \frac{P_{max} - P_{min}}{F_{sr}} = \frac{110 - 35}{12.5} = 6.0 \text{ in.}^2$$

To satisfy Eq. (3-9), AISC Specification 1.8.4, the

$$\text{required } r_{min} \geq \frac{L}{240} = \frac{12(12)}{240} = 0.6$$

From the AISC *Manual* the lightest WT section we can use is a WT7 × 21.5 or WT10.5 × 22.

Example 3-10   *Bolted Double-Angle Tension Member: Fatigue*

Design a double-angle tension member 20 ft long to support a dead load of 40 k and a live load that ranges from 65 k tension to 35 k compression with 600,000 loading cycles. Use A36 steel. Assume that the angle thickness is $\frac{3}{8}$ in., the long legs are back to back with a separation of $\frac{3}{8}$ in., and a single line of $\frac{7}{8}$-in.-diameter bolts is used to connect the member to the gusset plate. Also assume that there are at least three bolts in the line and stress category B.

Solution.   For A36 steel: $F_y = 36$ ksi, $F_u = 58$ ksi.

Loading range

$$
\begin{array}{lr}
\text{Dead load} & = \quad 40 \text{ k} \\
\text{Tensile live load} & = \quad \underline{65 \text{ k}} \\
P_{max} & = \quad 105 \text{ k} \\
\text{Dead load} & = \quad 40 \text{ k} \\
\text{Compressive live load} & = \underline{-35 \text{ k}} \\
P_{min} & = \quad 5 \text{ k}
\end{array}
$$

Allowable stress range. The following values are from the AISC *Specifications*, Appendix B.

1. Loading condition 3 for 600,000 cycles from Table B1
2. Stress category B from Table B2
3. The allowable stress range $F_{sr} = 18$ ksi from Table B3

Required properties for member selection. From Eqs. (3-5) and (3-10), we have

$$\text{required } A_g \geq \frac{P_{max}}{0.6F_y} = \frac{105}{22} = 4.77 \text{ in.}^2$$

$$\text{required } A_g \geq \frac{P_{max} - P_{min}}{F_{sr}} = \frac{105 - 5}{18} = 5.56 \text{ in.}^2$$

To satisfy Eq. (3-9), AISC Specification 1.8.4, the

$$\text{required } r_{\min} \geq \frac{L}{240} = \frac{20(12)}{240} = 1.0$$

From Eq. (3-6), the

$$\text{required } A_e \geq \frac{P_{\max}}{0.5F_u} = \frac{105}{29} = 3.621 \text{ in.}^2$$

The effective net area $A_e = 0.85 A_n$ from type of member (d) of Table 3-1, AISC Specification 1.14.2.2. Therefore,

$$\text{required } A_n \geq \frac{A_e}{0.85} = \frac{3.621}{0.85} = 4.259 \text{ in.}^2$$

The gross area from Eq. (3-9) will depend on the area of two standard holes in the $\frac{3}{8}$-in.-thick angles. The cross-sectional area of two holes is given by $A_h = n(d + \frac{1}{8})t = 2(\frac{7}{8} + \frac{1}{8})\frac{3}{8} = 0.75 \text{ in.}^2$. Therefore, the

$$\text{required } A_g \geq 4.259 + 0.75 = 5.01 \text{ in.}^2$$

The control value for $A_g = 5.56 \text{ in.}^2$. From the AISC *Manual* the lightest double angles we can use are two L4 × 4 × $\frac{3}{8}$ with equal legs or two L5 × 3 × $\frac{3}{8}$ with unequal legs.

PROBLEMS

Use the AISC *Specifications* for the following problems. Assume that all bolt holes are standard. In design determine the lightest section.

3-1. Determine the net area of a PL$\frac{3}{4}$ × 12 tension member with two rows of $\frac{3}{4}$-in. bolts parallel to the direction of the load.

3-2. Determine the net area of an L6 × 4 × $\frac{1}{2}$ tension member with two rows of $\frac{7}{8}$-in.-diameter bolts in the long leg.

3-3. Determine the net area of two L6 × 3$\frac{1}{2}$ × $\frac{3}{8}$ tension member placed with long legs back to back and connected at their ends to a $\frac{3}{8}$-in. gusset plate by two rows of $\frac{3}{4}$-in.-diameter bolts.

3-4. Determine the net area of a W14 × 90 tension member with two lines of 1-in. bolts in each flange. (There are four holes in any cross section.)

3-5. Determine the net area for the PL$\frac{3}{4}$ × 12 tension member shown. The holes are for 1-in.-diameter bolts and the stagger $s = 2\frac{1}{2}$ in.

3-6. What minimum stagger s may be used in Prob. 3-5 so that plate strength is controlled by the net section through two holes?

3-7. Determine the net area for the two L7 × 4 × $\frac{3}{8}$ tension members shown. The holes are for $\frac{7}{8}$-in.-diameter bolts and the stagger s is $2\frac{1}{2}$ in.

3-8. What minimum stagger s may be used in Prob. 3-7 so that angle strength will be controlled by the net section through four holes?

3-9. Determine the net area for the C12 × 30 tension member shown. The holes are for 1-in.-diameter bolts and the stagger $s = 3$ in.

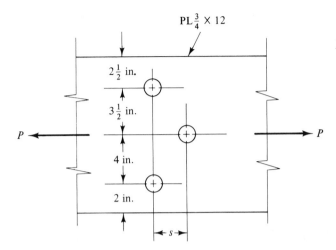

Holes for a 1-in.-diameter bolts

Prob. 3-5

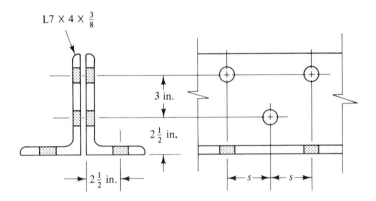

Holes for $\frac{7}{8}$-in.-diameter bolts

Prob. 3-7

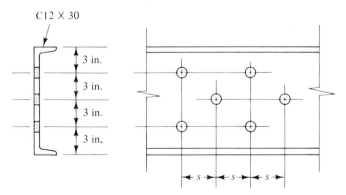

Holes for 1-in.-diameter bolts

Prob. 3-9

3-10. What minimum stagger s may be used in Prob. 3-9 so that channel strength will be controlled by the net section through two holes?

3-11. Determine the net effective area for the following W sections used as a tension member. There are two lines of 1-in.-diameter bolts in each flange and at least three bolts in each line. Assume no stagger of bolts in the lines. (There are four bolts in any cross section.)
   (a) W10 × 45.
   (b) W12 × 65.

3-12. Determine the slenderness ratio for a tension member with cross section and length as follows:
   (a) L5 × 5 × $\frac{1}{2}$, $L = 10$ ft 6 in.
   (b) Two L7 × 4 × $\frac{3}{4}$ (long legs back to back and separated by $\frac{3}{8}$ in.), $L = 15$ ft 9 in.
   (c) W12 × 50, $L = 33$ ft 7 in.
   (d) S10 × 35, $L = 14$ ft 9 in.

3-13. Determine the slenderness ratio for a tension member with cross section and length as shown.

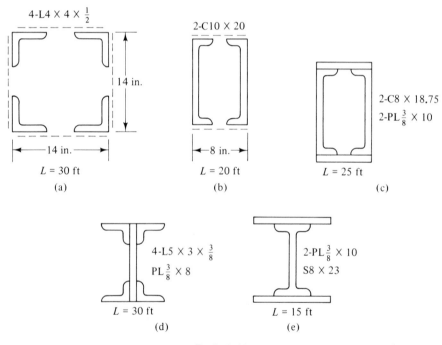

Prob. 3-13

3-14. Two $\frac{3}{4}$ × 10 plates are spliced together by cover plates as shown. If A36 steel is used and $s = 3$ in., determine the allowable tensile load.

3-15. Repeat Prob. 3-14 using A572 Grade 50 steel.

3-16. What minimum stagger s may be used in Prob. 3-14 so that plate strength is controlled by the net section through two holes?

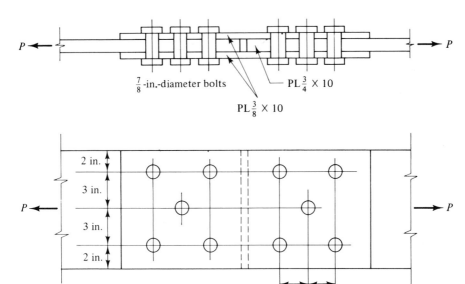

$\frac{7}{8}$-in.-diameter bolts   PL$\frac{3}{4}$ × 10

PL$\frac{3}{8}$ × 10

Prob. 3-14

3-17. A 15-ft-long L6 × 4 × t tension member supports a dead load of 44 k and a live load of 66 k. There are two lines of $\frac{7}{8}$-in. bolts in the long leg with at least three bolts in each line. Assume no stagger of bolts between the lines. If A36 steel is used, what angle thickness is required?

3-18. Repeat Prob. 3-17 using A572 Grade 50 steel.

3-19. Design member *CD* of the cantilever truss shown if $P = 30$ k. The member is a double-angle tension member. The legs are separated by $\frac{3}{8}$ in. and connected to a gusset plate with two lines of $\frac{3}{4}$-in.-diameter bolts with at least three bolts in each line. There is no stagger between the lines. Use A36 steel.

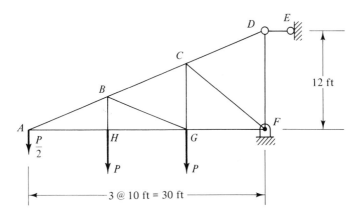

Prob. 3-19

3-20. Repeat Prob. 3-19 using $\frac{7}{8}$-in.-diameter bolts, A572 Grade 50 steel, and $P = 40$ k.

3-21. Design a double-angle tension member 20 ft long to support a dead load of 60 k and a live load of 90 k. The member supports floor beams, and impact must be considered. The long legs are back to back and are separated by $\frac{3}{8}$ in. and connected to a gusset plate. Assume two lines of $\frac{7}{8}$-in. bolts in the long legs with at least three bolts in each line. There is no stagger of bolts between the lines. Use A36 steel.

3-22. Repeat Prob. 3-21 using A572 Grade 50 steel.

3-23. One edge of a balcony floor is supported by a hanger. The hanger is 12 ft long and supports a dead load of 5.4 k and a live load of 27 k. Assume that a single row of at least three $\frac{3}{4}$-in.-diameter bolts connects the hanger to the balcony floor. Using A36 steel, select the lightest single-angle tension member that will safely support the floor. Impact must be considered.

3-24. Repeat Prob. 3-23 using a double-angle tension member if the legs are separated by $\frac{3}{8}$ in.

3-25. Select a W10 section tension member 30 ft long to support a dead load of 90 k, a live load of 130 k, and a wind load of 145 k. There are two lines of 1-in.-diameter bolts in each flange and at least three holes in each line. Assume no stagger of bolts between the lines. (There are four holes in any cross section.) Use A36 steel.

3-26. Repeat Prob. 3-25 using a W8 section.

3-27. Design a WT8 section tension member 20 ft long to support a dead load of 65 k, a live load of 90 k, and a possible wind load of 40 k. Assume welded connections to the flanges so that there are no holes. Use A36 steel.

3-28. Repeat Prob. 3-27 using a WT6 section and A572 Grade 60 steel.

3-29. Select a C12 section tension member 10 ft long to support a dead load of 50 k and a live load of 80 k. The member is connected to a gusset plate as shown. Use $\frac{3}{4}$-in. bolts and A36 steel.

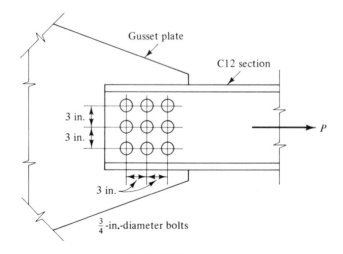

Prob. 3-29

3-30. Repeat Prob. 3-29 using a C10 section and A572 Grade 50 steel.

3-31. Design a WT section 10 ft long to support a dead load of 35 k and a live load that ranges from 50 k tension to 8 k compression with 100,000 loading cycles. Use A36 steel. The fillet welded end connection is similar to case 17 in Fig. B1 of Appendix B of the AISC *Manual.*

3-32. Select a double-angle tension member 15 ft long to support a dead load of 50 k and a live load that ranges from 80 k tension to 40 k compression with 550,000 loading cycles. Use an angle thickness of $\frac{1}{2}$ in. with the long legs back to back with a separation of $\frac{3}{8}$ in. Assume that there are two lines of 1-in.-diameter bolts with at least three bolts in each line and we have stress category B. Use A36 steel.

4

COMPRESSION MEMBERS
Columns

4-1 INTRODUCTION

Members that are subject to axial compressive loads are usually called *columns*. Compression members in a truss may be known as *chord members* or *web members*, depending on their truss location. Columns may also be called by various other names, such as *braces* or *struts*. *Knee braces* are short compression members that are used between the columns and a roof truss or beam of a building frame. *Eave struts* are short compression members that connect adjacent building frames braced at the eaves. In each example, the member described supports a compressive load and will be called a *column* or *compression member*.

4-2 TYPES OF COMPRESSION FAILURE

There are fundamental differences in the way tension and compression members or columns fail. The tension member remains straight under a tensile load and failure may occur either by yielding of the gross section or fracture of the net section. Two different types of failure must be considered for the column. Failure depends on the length of the member compared to its cross-sectional dimensions. In a test to failure an axial compressive load applied to a 1-in.-diameter steel rod 2 in. long causes *yielding* or crushing of the rod. A steel rod with the same diameter 20 in. long fails by *buckling* or sudden bending and collapse. Thus we see that for *short columns* the failure load is characterized by yielding and for *long columns* by buckling or instability. Members with lengths that fall between short and long columns are called *intermediate columns.* They fail by both yielding and buckling.

4-3 SECTIONS USED FOR STEEL COLUMNS

Compression members commonly consist of pipes or round tubing, square or rectangular tubing, and such shapes as angles, channels, tees, or W shapes. They may also be built up of rolled shapes and plates. The kind of compression member to be used depends on the end connections and type of structure.

Cross sections for a few of the many possible compression members are shown in Fig. 4-1. The pipe or round tubing shown in (a) may be used for columns in one-story warehouses, supermarkets, and residential basements and garages. Wide-flange shapes (b) are commonly used as building columns, and structural tubes (c) are finding increased favor as building columns because

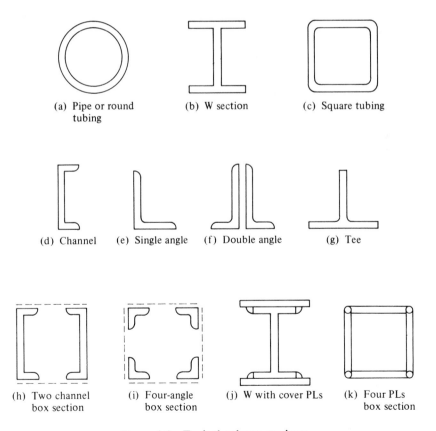

(a) Pipe or round   (b) W section   (c) Square tubing
     tubing

(d) Channel   (e) Single angle   (f) Double angle   (g) Tee

(h) Two channel   (i) Four-angle   (j) W with cover PLs   (k) Four PLs
     box section   box section   box section

Figure 4-1 Typical column sections.

of the use of welded connections. Single channels (d) can be used as compression members if intermediate bracing is provided in the weak direction. Angle sections and tees (e), (f), and (g) are used mainly as bracing members and compression members in trusses. The built-up sections (h), (i), (j), and (k) may be used as building columns or web members in large trusses.

4-4 THE EULER COLUMN FORMULA

The mode of failure for a long column is buckling or sudden bending and collapse. The smallest column load that produces buckling is called the *critical* or *Euler load*. For a load less than the critical load, the column remains straight and in stable equilibrium. For a load equal to or greater than the critical, the column is unstable and fails. The critical load for a long column that has pin supports at both ends is given by the *Euler column formula*

$$P_c = \frac{\pi^2 EI}{L^2} \qquad (4\text{-}1)$$

where P_c = critical load, k
   E = modulus of elasticity, ksi
   I = least moment of inertia of the constant cross-sectional area of the column, in.^{4}
   L = length of the column, in.
The Euler column formula was derived by the Swiss mathematician Leonhard Euler in 1757. (For a derivation, see any strength-of-materials textbook.)

4-5 EFFECTIVE BUCKLING LENGTH OF COLUMNS

For end conditions other than pin supports, the shape of the deflection curve can be used to modify the Euler formula. In Fig. 4-2 we show the deflection curves for columns with various end conditions. In each case, the length of a single loop of a sine curve represents the effective buckling length L_e of the column. The extent of the loop is marked on each of the deflection curves. The effective length can be expressed as $L_e = KL$, where K is an effective length factor and L is the actual length of the column. The factor K depends on end conditions. Substituting the effective length into Eq. (4-1), the critical load becomes

$$P_c = \frac{\pi^2 EI}{(KL)^2} \qquad (4\text{-}2)$$

For example, with both ends of the column fixed as shown in Fig. 4-2(b), $K = 0.5$ and the critical load $P_c = 4\pi^2 EI/L^2$. Thus we see that the critical load for a column with fixed ends is four times the critical load for an identical column with both ends pinned.

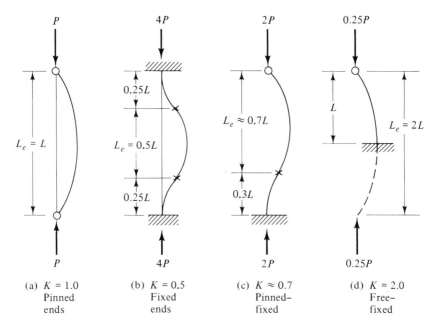

Figure 4-2   Effective length of columns with different end conditions.

4-6 FURTHER COMMENTS ON THE EULER COLUMN FORMULA

It will be useful to express the Euler formula in a different form. By definition, $I = Ar^2$, where I is the moment of inertia of the cross-sectional area of the column, A is the cross-sectional area of the column, and r is the radius of gyration. Substituting for the moment of inertia in Eq. (4-2) we have another form of the Euler column formula,

$$P_c = \frac{\pi^2 EI}{(KL)^2} = \frac{\pi^2 EAr^2}{(KL)^2}$$

or in terms of the critical stress,

$$F_c = \frac{P_c}{A} = \frac{\pi^2 E}{(KL/r)^2} \qquad (4\text{-}3)$$

where F_c = critical stress, ksi

E = modulus of elasticity, ksi

KL/r = slenderness ratio—the effective length of the column divided by the minimum radius of gyration of the cross-sectional area of the column, both in inches

Notice that the critical load or stress as given by Eq. (4-2) or (4-3) depends on the modulus of elasticity E, a measure of the stiffness of the material. The modulus is essentially a constant for all steels. Therefore, the load-supporting capacity of a long steel column made of A36 steel with a yield stress $F_y = 36$ ksi is the same as one made of A572 steel with a yield stress $F_y = 50$ ksi.

Because the Euler column formula is based on Hooke's law, the formula is not valid for stresses above the proportional limit of the material. In fact, we are now able to define what is meant by a long column. A *long column* is one in which the critical stress F_c is less than the proportional limit f_p for the material.

Steel in the following examples has a modulus of elasticity $E = 29(10^3)$ ksi, a yield stress $F_y = 36$ ksi, and a proportional limit $F_p = 34$ ksi.

Example 4-1   *W Section Columns: Critical Load*

Determine the critical load for a 25 ft long W8 × 35 steel column with one end pinned and the other end fixed.

Solution.   Properties of the W8 × 35 from the AISC *Manual*: $A = 10.3$ in.^{2}, $r_x = 3.51$ in., $r_y = 2.03$ in. For one end pinned and the other end fixed, $K = 0.7$ [Fig. 4-2(c)]. The *minimum* radius of gyration $r_y = 2.03$ in. and the *maximum* slenderness ratio

$$\frac{KL}{r_y} = \frac{0.7(25)(12)}{2.03} = 103.4$$

From the Euler column formula [Eq. (4-3)] we have the critical stress

$$F_c = \frac{\pi^2 E}{(KL/r)^2} = \frac{\pi^2(29,000)}{(103.4)^2} = 26.7 \text{ ksi} < F_p = 34 \text{ ksi}$$

The critical load is given by the product of the critical stress and gross cross-sectional area. Thus

$$P_c = F_c A = 26.7(10.3) = 275 \text{ K}$$

Example 4-2   *Double-Channel Column Member: Critical Load*

Determine the critical load for a 35-ft-long built-up column with cross section as shown in Fig. 4-3. Both ends are pinned.

Solution.   Properties of one C12 × 30 from the AISC *Manual*: $A = 8.82$ in.^{2}, $r_{xc} = 4.29$ in., $r_{yc} = 0.763$ in., $x = 0.674$ in. The radius of gyration about the x axis for the two channels is the same as it is for one channel. Thus $r_x = r_{xc} = 4.29$ in. The radius of gyration with respect to the y axis must be found from the parallel axis theorem. The distance between the two parallel axes y_c and y [Fig. 4-3(b)] is

$$d_x = \frac{h}{2} - x = 7 - 0.674 = 6.326 \text{ in.}$$

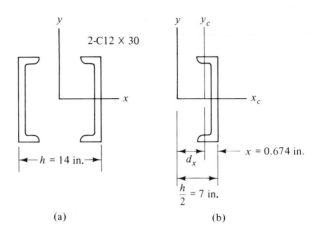

Figure 4-3   Example 4-2.

The parallel axis theorem for the radius of gyration is given by

$$r_y^2 = r_{yc}^2 + d_x^2$$

Therefore,

$$r_y^2 = (0.763)^2 + (6.326)^2 = 40.6$$

$$r_y = 6.37 \text{ in.}$$

The minimum radius of gyration gives the maximum slenderness ratio. Thus

$$\frac{KL}{r_x} = \frac{1.0(35)(12)}{4.29} = 97.9$$

From Eq. (4-3), the critical stress

$$F_c = \frac{\pi^2 E}{(KL/r)^2} = \frac{\pi^2(29,000)}{(97.9)^2} = 29.9 \text{ ksi}$$

For two channels the critical load is given by

$$P_c = F_c A = 2(29.9)(8.82) = 527 \text{ k}$$

4-7 SHORT, INTERMEDIATE, AND LONG COLUMNS

To illustrate the difference between short, intermediate, and long columns we display the stress–strain and the critical stress versus slenderness ratio curves side by side in Fig. 4-4. From the curves we make the following observations.

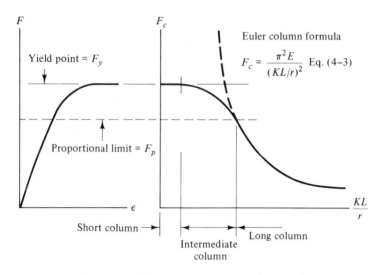

Figure 4-4   Critical or failure stress for short, intermediate, and long columns.

Long columns.   The critical stress F_c for a long column is less than the proportional limit F_p of the material. Therefore, a long column buckles *elastically.*

Short columns.   The failure stress for a short column is equal to the yield stress F_y and no buckling occurs.

Intermediate columns.   The failure stress for an intermediate column falls between the proportional limit F_p and the yield stress F_y. The member fails by both yielding and buckling. Thus buckling of an intermediate column is said to be *inelastic.*

4-8 AISC COLUMN DESIGN EQUATIONS

The AISC *Specifications* recommend the Euler formula for long columns and a parabolic formula for short and intermediate columns, as shown in Fig. 4-5. Factors of safety are applied to obtain the allowable stresses.

Based on column tests as well as the measurement of residual stresses in rolled steel columns, the AISC *Specifications* assume that the proportional limit occurs at a stress equal to one-half of the yield stress. Therefore, the parabola extends from a stress equal to the yield stress, where $KL/r = 0$, to a stress equal to one-half the yield stress, where $KL/r = C_c$. Because the parabola joins the Euler curve at $KL/r = C_c$, we can evalute C_c by equating

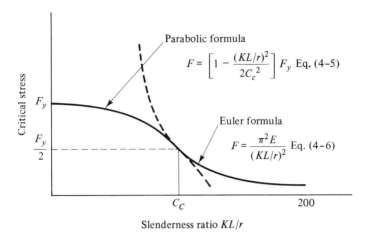

Figure 4-5 AISC column formulas without factors of safety.

one-half the yield stress to the stress given by the Euler formula. That is,

$$\frac{F_y}{2} = \frac{\pi^2 E}{(C_c)^2} \quad \text{or} \quad C_c = \sqrt{\frac{2\pi^2 E}{F_y}} \tag{4-4}$$

The parabolic equation is given by

$$F_c = \frac{P}{A} = \left[1 - \frac{(KL/r)^2}{2C_c^2} \right] F_y \quad \text{where } 0 \le \frac{KL}{r} \le C_c \tag{4-5}$$

and the Euler equation is given by

$$F_c = \frac{P}{A} = \frac{\pi^2 E}{(KL/r)^2} \quad \text{where } C_c \le \frac{KL}{r} \le 200 \tag{4-6}$$

Equation (4-5) is divided by a variable factor of safety, defined by

$$\text{F.S.} = \frac{5}{3} + \frac{3(KL/r)}{8C_c} - \frac{(KL/r)^3}{8C_c^3} \tag{4-7}$$

to give the allowable stress. (This factor of safety has a value of 1.67 at $KL/r = 0$ and a value of 1.92 at $KL/r = C_c$.) Equation (4-6) is divided by a constant factor of safety $\frac{23}{12} = 1.92$ to give the allowable stress. With the modulus of elasticity for structural steel taken as $E = 29(10^3)$ ksi, the allowable stress equations [AISC Formulas (1.5-1) and (1.5-2)] for column design are given by

$$F_a = \frac{\left[1 - \frac{(KL/r)^2}{2C_c^2} \right] F_y}{\dfrac{5}{3} + \dfrac{3(KL/r)}{8C_c} - \dfrac{(KL/r)^3}{8C_c^3}} \quad \text{where } 0 \le \frac{KL}{r} \le C_c \tag{4-8}$$

and

$$F_a = \frac{12\pi^2 E}{23(KL/r)^2} = \frac{149,000}{(KL/r)^2} \text{ where } C_c \le \frac{KL}{r} \le 200 \qquad (4\text{-}9)$$

Equations (4-8) and (4-9) are applicable to all main members and secondary members where L/r is equal to or less than 120. For bracing and secondary members having L/r exceeding 120, the allowable stress is increased by AISC Formula (1.5-3). That is,

$$F_a = \frac{\text{Eq. (4-8) or (4-9)}}{1.6 - L/200r} \qquad (4\text{-}10)$$

Notice that the factor K is taken as unity.

The length KL used in the AISC column formulas is defined in Sec. 4-5 as the effective buckling length of the column. Table C1.8.1 of AISC Commentary (Sec. 1.8) gives theoretical and recommended values of K. Table 4-1 in the text is adapted from that source.

For most column design it is difficult to evaluate the degree of fixity that exists at the ends of the columns. Columns are very often members of frames and their effective lengths are dependent on other members of the frame or on bracing of the frame. If the frame is braced, no sidesway (motion of the top of the column relative to the bottom) is possible. The buckled shape of such columns resembles cases (a), (b), and (c) of Table 4-1. For buckling of an unbraced frame the top of the columns move relative to the bottom and sidesway occurs. Cases (d), (e), and (f) are examples of sidesway buckling. Sidesway buckling is discussed in Chapter 6 on beam-columns.

Example 4-3 *Allowable Load for a W Section Column*

Determine the allowable load for a W8 × 67 column made of A36 steel with a length of 17 ft. One end is pinned and the other end fixed.

Solution. Properties of the W8 × 67 are from the AISC *Manual* and K is from Table 4-1: $A = 19.7$ in.^{2}, $r_x = 3.72$ in., $r_y = 2.12$ in., and the recommended effective length $K = 0.8$. The controlling slenderness ratio

$$\frac{KL}{r_y} = \frac{0.8(17)(12)}{2.12} = 76.98$$

From Eq. (4-4),

$$C_c = \sqrt{\frac{2\pi^2 E}{F_y}} = \sqrt{\frac{2\pi^2(29,000)}{36}} = 126.1$$

Therefore,

$$\frac{KL}{r_y} = 76.98 < C_c = 126.1$$

TABLE 4-1   Effective Lengths for Main Compression Members

Buckled shape of column is shown by dashed line	(a)	(b)	(c)	(d)	(e)	(f)
Theoretical K value	1.0	0.7	0.5	2.0	2.0	1.0
Recommended design value when ideal conditions are approximated	1.0	0.80	0.65	2.10	2.0	1.2

End condition code	
	Rotation free and translation fixed
	Rotation fixed and translation fixed
○	Rotation free and translation free
▨	Rotation fixed and translation free

Source: Adapted from Table C1.8.1 of the Commentary on the AISC Specification. (Courtesy of the American Institute of Steel Construction, Inc.)

Use Eq. (4-8).

$$\frac{KL/r}{C_c} = \frac{76.98}{126.1} = 0.6105$$

$$F_a = \frac{\left[1 - \frac{1}{2}\left(\frac{KL/r}{C_c}\right)^2\right]F_y}{\frac{5}{3} + \frac{3}{8}\left(\frac{KL/r}{C_c}\right) - \frac{1}{8}\left(\frac{KL/r}{C_c}\right)^3}$$

$$F_a = \frac{[1 - \frac{1}{2}(0.6105)^2]36}{\frac{5}{3} + \frac{3}{8}(0.6105) - \frac{1}{8}(0.6105)^3} = 15.69 \text{ ksi}$$

Thus

$$P_a = F_a A = 15.69(19.7) = 309 \text{ k}$$

Example 4-4 *Allowable Load for a Built-up Column*

Determine the allowable load for a 23-ft column made of A36 steel. One end of the column is fixed and the other end free. The cross section is shown in Fig. 4-6.

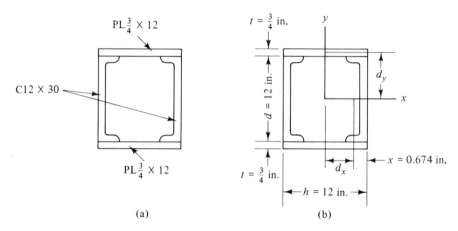

(a)   (b)

Figure 4-6 Example 4-4.

Solution. Values of the radius of gyration must be calculated for the built-up cross section. Properties of the C12 × 30 are from the AISC *Manual* and are shown in Table (a). Transfer distances for the channel and plate are shown in Fig. 4-6(b). For the plate

$$I_{xc} = \frac{bh^3}{12} = \frac{(12)(0.75)^3}{12} = 0.4 \text{ in.}^4$$

$$I_{yc} = \frac{hb^3}{12} = \frac{(0.75)(12)^3}{12} = 108 \text{ in.}^4$$

The transfer distances for the channel on the right (R)

$$d_x = \frac{h}{2} - x = 6 - 0.674 = 5.326 \text{ in.}$$

and for the plate on the top (T)

$$d_y = \frac{d + t}{2} = \frac{12 + 0.75}{2} = 6.375 \text{ in.}$$

TABLE (a) for Example 4-4

Shape	A	d_x	d_y	I_{xc}	Ad_y^2	I_{yc}	Ad_x^2
C12 × 30 (R)	8.82	5.326	0	162	0	5.14	250.2
C12 × 30 (L)	8.82	5.326	0	162	0	5.14	250.2
PL$\frac{3}{4}$ × 12 (T)	9.0	0	6.375	—	365.8	108.0	0
PL$\frac{3}{4}$ × 12 (B)	9.0	0	6.375	—	365.8	108.0	0
	35.64 (in.^{2})	—	—	324 (in.^{4})	731.6 (in.^{4})	226.3 (in.^{4})	500.4 (in.^{4})

The results are tabulated as shown. From the parallel axis theorem and the sums in Table (a),

$$I_x = I_{xc} + (Ad_y^2) = 324 + 731.6$$

$$= 1055.6 \text{ in.}^4$$

$$I_y = I_{yc} + (Ad_x^2) = 226.3 + 500.4$$

$$= 726.7 \text{ in.}^4$$

The radiuses of gyration are given by

$$r_x = \sqrt{\frac{I_x}{A}} = \sqrt{\frac{1055.6}{35.64}} = 5.44 \text{ in.}$$

and

$$r_y = \sqrt{\frac{I_y}{A}} = \sqrt{\frac{726.7}{35.64}} = 4.51 \text{ in.}$$

The effective length factor $K = 2.1$ from Table 4-1. Therefore, the controlling slenderness ratio is given by

$$\frac{KL}{r_y} = \frac{2.1(23)(12)}{4.51} = 128.5$$

From Eq. (4-4),

$$C_c = \sqrt{\frac{2\pi^2 E}{F_y}} = \sqrt{\frac{2\pi^2(29,000)}{36}} = 126.1$$

The slenderness ratio is larger than C_c; therefore, Eq. (4-9) is used.

$$F_a = \frac{12\pi^2 E}{23(KL/r)^2} = \frac{12\pi^2(29,000)}{23(128.5)^2}$$

$$= 9.04 \text{ ksi}$$

The allowable load

$$P_a = F_a A = 9.04(35.64) = 322 \text{ k}$$

Example 4-5   *Column Load Tables for W Section Column*

Rework Example 4-3 using the column load tables in the AISC *Manual.*

Solution. The W8 × 67 has an effective length $KL = 0.8(17) = 13.6$ ft. From the tables, the capacity of the column is given by interpolation.

$$P_a = 316 - \frac{13.6 - 13}{14 - 13}(316 - 304)$$

$$= 308.8 = 390 \text{ k}$$

Column load tables are available for various rolled shapes.

Example 4-6   *Allowable Stress Tables for Built-up Columns*

Rework Example 4-4 using Table 3-36 of the AISC *Manual.*

Solution. For the built-up section shown in Fig. 4-6, we must first determine the controlling slenderness ratio and cross-sectional area as in Example 4-4. For a slenderness ratio of 128.5 we interpolate from the AISC *Manual*, Table 3-36, to find the allowable stress.

$$F_a = 9.11 - \frac{128.5 - 128}{129 - 128}(9.11 - 8.97)$$

$$= 9.04 \text{ ksi}$$

The allowable load is given by

$$P_a = F_a A = 9.04(35.64) = 322 \text{ k}$$

There are allowable stress tables for columns with yield stresses of 36 ksi and 50 ksi, a range of slenderness ratios from 1 to 200, and including main and secondary members in the AISC *Manual.*

Example 4-7   *Braced W Section Column*

A pin-ended W12 × 65 column made of A36 steel is supported at midheight perpendicular to the y axis of its cross section as shown in Fig. 4-7. Determine the allowable axial load if the member has a length of 20 ft 6 in.

Solution. Properties of the W12 × 65: $A = 19.1$ in.^{2}, $r_x = 5.28$ in., $r_y = 3.02$ in. The column is supported at midheight perpendicular to the y axis. Therefore, the effective length for buckling about the y axis $K_y L_y$ is equal to one-half the column length and the effective length for buckling about the x axis $K_x L_x$ is equal to the column length. The slenderness ratios are given by

$$\frac{K_x L_x}{r_x} = \frac{20.5(12)}{5.28} = 46.6$$

$$\frac{K_y L_y}{r_y} = \frac{10.25(12)}{3.02} = 40.7$$

The maximum slenderness ratio 46.6 is used to determine the allowable stress. From Table 3-36 of the AISC *Manual* the allowable stress F_a, by interpolation, is 18.65 ksi. The allowable load is given by

$$P_a = F_a A = 18.65(19.1) = 356 \text{ k}$$

Alternative Solution. The column load tables are based on buckling about the y axis. In this problem the controlling slenderness ratio involves buckling about the x axis. To use the tables we determine an effective length based on the y axis. Thus

$$\frac{(K_y L_y)_{\text{eff}}}{r_y} = \frac{K_x L_x}{r_x} \qquad \text{or} \qquad (K_y L_y)_{\text{eff}} = \frac{r_y}{r_x} K_x L_x$$

Using the ratio $r_x / r_y = 1.75$ tabulated for the W12 × 65 section, the effective length

$$\frac{r_y}{r_x} K_x L_x = \frac{20.5}{1.75} = 11.7 \text{ ft}$$

The allowable load by interpolation for a column 11.7 ft long from the column

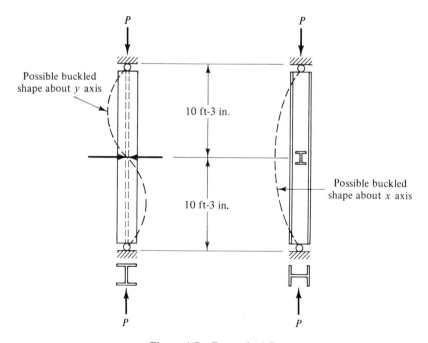

Figure 4-7   Example 4-7.

load table is

$$P_a = 356 \text{ k}$$

The column load tables are based on buckling about the y axis. Therefore, the method outlined in this example is used only when the controlling slenderness ratio involves buckling about the x axis.

Example 4-8 *Secondary Member with a Slenderness Ratio Greater than 120*

Determine the allowable load for a W4 × 13 column used as a bracing member. The column is made of A36 steel and has an actual length of 12 ft.

Solution.   Properties of the W4 × 13: $A = 3.83 \text{ in.}^2$, $r_x = 1.72$ in., $r_y = 1.00$ in. Bracing members are classified as secondary members and K is taken as 1.0 (AISC Specification 1.5.1.3.3). Thus the slenderness ratio is

$$\frac{KL}{r_y} = \frac{1(12)(12)}{1.00} = 144$$

The column load tables give loads for main members and secondary members whose slenderness ratio is 120 or less. Therefore, we use an alternative approach for this problem. From the AISC *Manual*, Table 3-36, the allowable stress for a secondary member with a slenderness ratio of 144 is $F_a = 8.18$ ksi. Thus the allowable load is given by

$$P_a = F_a A = 8.18(3.83) = 31.3 \text{ k}$$

4-9 DESIGN OF AXIALLY LOADED COLUMNS

The design of axially loaded columns is based on Eqs. (4-8) and (4-9). Since the allowable stress is unknown before the cross section of the column is selected, column design without column load tables requires a trial-and-error or indirect process. We summarize the design process as follows:

1. Assume in allowable stress F_a.
2. Based on the allowable stress, calculate the required area.
3. Select a trial section to satisfy the required area.
4. Determine the actual allowable load for the trial section.
5. Compare the actual allowable load and the required load.
   a. If the actual allowable load is greater than the required load, select a lighter trial section and repeat steps 4 and 5.
   b. If the actual allowable load is only slightly larger than the required load, the design is complete. Use the trial section for the design.
   c. If the actual allowable load is less than the required load, select a heavier section and repeat steps 4 and 5.

The outlined design method is illustrated in the following example.

Example 4-9 *Trial-and-Error Design of W Section Column*

Select a W12 column section to support a load of 300 k. The member is 35 ft long and is fixed at the bottom and pinned at the top. Use A36 steel.

Solution.   From Table 4-1 the recommended $K = 0.80$.

1. Assume the allowable stress $F_a = 13$ ksi.
2. The required area is given by

$$A = \frac{P}{F_a} = \frac{300}{13} = 23 \text{ in.}^2$$

3. Try W12 × 79: $A = 23.2$ in.^{2}, $r_x = 5.34$ in., $r_y = 3.05$ in.
4. The maximum slenderness ratio

$$\frac{KL}{r_y} = \frac{0.8(35)(12)}{3.05} = 110.2$$

From the AISC *Manual*, Table 3-36, by interpolation $F_a = 11.64$ ksi. The allowable load is

$$P_a = F_a A = 11.64(23.2) = 270 \text{ k}$$

5. Since the allowable load is less than the required load, a heavier section is requied.

Second Trial

3. Try W12 × 87: $A = 25.6$ in.^{2}, $r_x = 5.38$ in., $r_y = 3.07$ in.
4. The maximum slenderness ratio

$$\frac{KL}{r_y} = \frac{0.8(35)(12)}{3.07} = 109.4$$

From the AISC *Manual*, Table 3-36, by interpolation $F_a = 11.75$ ksi. The allowable load is

$$P_a = F_a A = 11.75(25.6) = 301 \text{ k}$$

Use the W12 × 87.

Alternative Method. From the column load table in the AISC *Manual* a W12 × 87 with an effective length of $KL = 0.8(35) = 28$ ft can support an allowable load $P_a = 301$ k. This method gives the same answer as before. The trial-and-error method is used for the design of columns that do not appear in the column load tables.

Example 4-10 *Design of a Double-Angle Compression Member: Wind Loads*

Select the lightest double-angle compression member made of A36 steel for the top chord of a truss as shown in Fig. 4-8. In the plane of the truss, cross

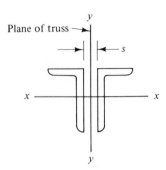

Figure 4-8   Example 4-10.

members provide bracing at intervals of $L_x = 5$ ft. For out-of-plane buckling, bracing is provided at intervals of $L_y = 10$ ft. The design load $P = 100$ k includes the wind load.

Solution.   The column load tables in the AISC *Manual* do not include the effect of wind. To include the effect of wind the loads given in the table should be increased by one-third (AISC Specification 1.5.6). Alternatively, the design load can be reduced by one-fourth. The allowable loads given in the column load tables can then be used directly to select the double-angle compression member. The reduced column load is given by

$$P_{\text{red}} = \frac{3}{4}P = \frac{3}{4}(100) = 75 \text{ k}$$

The effective lengths with respect to the x and y axes are $K_x L_x = 1(5) = 5$ ft and $K_y L_y = 1(10) = 10$ ft. From the column load table in the AISC *Manual* two $L3\frac{1}{2} \times 3\frac{1}{2} \times \frac{3}{8}$ back to back and separated by $\frac{3}{8}$ in. can support 88 k with respect to the x axis and 79 k with respect to the y axis. Two $L3\frac{1}{2} \times 3 \times \frac{3}{8}$ with short legs back to back and separated by $\frac{3}{8}$ in. can support 76 k with respect to the x axis and 75 k with respect to the y axis. Use two $L3\frac{1}{2} \times 3 \times \frac{3}{8}$.

Example 4-11   *Design of a Braced W Section Column*

Select the most economical W14 column made of A36 steel with pinned ends if the column is braced at midheight perpendicular to the y axis of the cross section. The column load and lengths are as follows:
   (a) $P = 500$ k, $L = 24$ ft.
   (b) $P = 300$ k, $L = 24$ ft.

Solution.   The AISC *Manual* column load tables are based on buckling about the y axis. Therefore, as indicated in Example 4-7, the two effective lengths that must be determined are

$$K_y L_y \quad \text{and} \quad \frac{r_y}{r_x} K_x L_x$$

The ratio r_y/r_x cannot be determined until a tentative section has been selected. We make the selection on the basis of $K_y L_y$ and check to see if it is still adequate based on $(r_y/r_x)K_x L_x$.

(a) For $P = 500$ k and $L = 24$ ft, $K_y L_y = 1(12) = 12$ ft. From the tables we select a W14 × 90 section with $r_x/r_y = 1.66$. The effective length with respect to the x axis $(r_y/r_x)K_x L_x = 1(24)/1.66 = 14.5$ ft. Thus the controlling effective length is 14.5 ft. The capacity of the W14 × 90 with an effective length of 14.5 ft is 493 k. Therefore, we must try a heavier section. From the table we select a W14 × 99. The W14 × 99 has a controlling effective length of 14.5 ft and a capacity of 542 k. Use a W14 × 99.

(b) For $P = 300$ k and $L = 24$ ft, $K_y L_y = 1(12) = 12$ ft. From the table we select a W14 × 61 with $r_x/r_y = 2.44$. The effective length with respect to the x axis $(r_y/r_x)K_x L_x = 1(24)/2.44 = 9.8$ ft. Therefore, the controlling effective length is 12 ft. The capacity of the W14 × 61 with an effective length of 12 ft is 314 k. Use a W14 × 61.

4-10 LOCAL BUCKLING

So far we have considered only buckling of the entire column. When the member is made up of thin elements, however, it is quite possible that buckling of one or more of the elements may occur at loads that are less than the buckling load for the column. Such buckling is called *local buckling*. Local buckling depends on the width-to-thickness ratio b/t of the compression elements that make up the column cross section and the edge support conditions

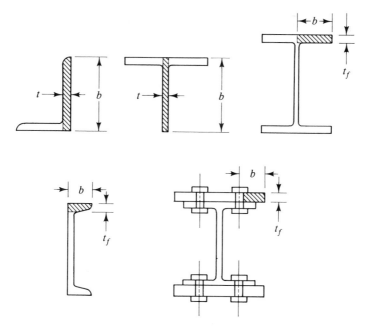

Figure 4-9   Unstiffened compression elements.

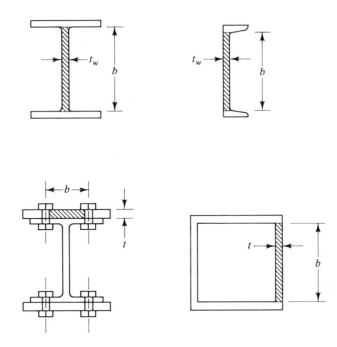

Figure 4-10   Stiffened compression elements.

for those elements. AISC Specification 1.9 places limitations on the ratios under two different categories: unstiffened elements and stiffened elements. If one edge of the element is supported and the other free, the element is *unstiffened*. If both edges are supported, the element is *stiffened*. (The edges in both categories are parallel to the direction of the compressive stress.) Examples are shown in Figs. 4-9 and 4-10.

To prevent local buckling of the compression elements before the column as a whole buckles, the allowable stress for the column is reduced and/or the effective area of the elements is reduced (see AISC *Specifications*, Appendix C).

4-11 LACED COLUMNS

The purpose of lacing bars, battens, and cover plates for built-up columns is to tie all the elements of the cross section together so that they act as a unit in supporting the column load. Examples are shown in Fig. 4-11.

Requirements for lacing are given in AISC Specification 1.18.2.6. If the critical stress is less for the section of a column between lacing connections *P* and *Q* shown in Fig. 4-12(a) than for the column as a whole, failure will occur due to local buckling (shown as a dashed line in the figure). Local failure can then lead to column failure [Fig. 4-12(b)]. To prevent this type of

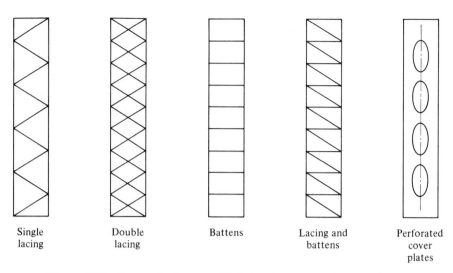

| Single lacing | Double lacing | Battens | Lacing and battens | Perforated cover plates |

Figure 4-11   Types of columns with lacing, battens, and cover plates.

failure, lacing must be spaced so that the L/r ratio of the individual elements between the lacing connections does not exceed the KL/r ratio for the entire built-up column.

Lacing is designed to support a shear force normal to the column axis equal to 2 percent of the total compressive force in the member. Lacing bars are considered secondary members and their slenderness ratios are limited to

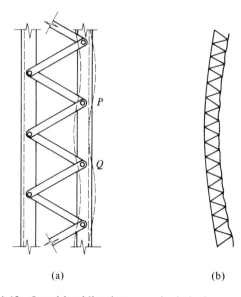

(a)   (b)

Figure 4-12   Local buckling between single lacing connections.

TABLE 4-2   Approximate Radiuses of Gyration for Various Shapes

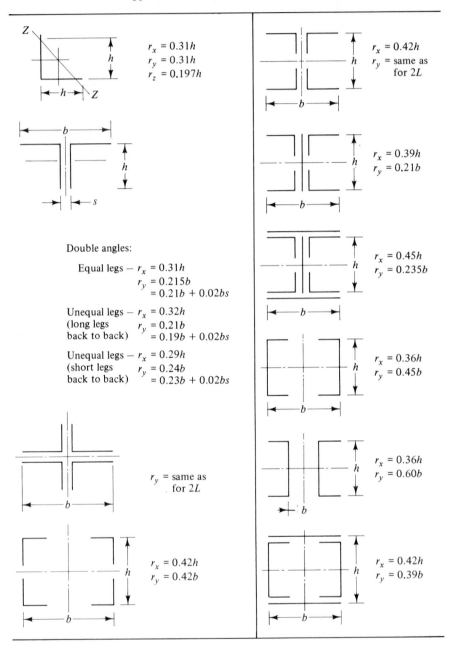

$r_x = 0.31h$
$r_y = 0.31h$
$r_z = 0.197h$

$r_x = 0.42h$
r_y = same as for 2L

$r_x = 0.39h$
$r_y = 0.21b$

Double angles:

Equal legs — $r_x = 0.31h$
$r_y = 0.215b$
$= 0.21b + 0.02bs$

$r_x = 0.45h$
$r_y = 0.235b$

Unequal legs — $r_x = 0.32h$
(long legs $r_y = 0.21b$
back to back) $= 0.19b + 0.02bs$

Unequal legs — $r_x = 0.29h$
(short legs $r_y = 0.24b$
back to back) $= 0.23b + 0.02bs$

$r_x = 0.36h$
$r_y = 0.45b$

r_y = same as for 2L

$r_x = 0.36h$
$r_y = 0.60b$

$r_x = 0.42h$
$r_y = 0.42b$

$r_x = 0.42h$
$r_y = 0.39b$

Source: Adapted from J.A.L. Waddell, *Bridge Engineering*, John Wiley & Sons, Inc., New York, 1925. (Courtesy of John Wiley & Sons, Inc.)

140 for single lacing and 200 for double lacing. The inclination of the lacing bars with the axis of the column should preferably be not less than 60° for single lacing and 45° for double lacing. If the distance between lines of fasteners or welds connecting the lacing is more than 15 in., the specifications require double lacing, or if single lacing is used, the lacing members must be angles.

In addition to lacing, compression members built up from plates or shapes must have tie plates at each end and at intermediate points if the lacing is interrupted. Requirements for tie plates are given in AISC Specification 1.18.2.5.

As an alternative to lacing and tie plates, AISC Specification 1.18.2.7 permits their function to be performed by continuous cover plates perforated by a succession of access holes. However, such columns would only be used where large axial loads and bending moments are acting together. The AISC *Specifications* do not consider battened columns.

To aid in the design of built-up columns, we show approximate radiuses of gyration in Table 4-2. By adjusting the dimensions of a column the radius of gyration of the cross section can be controlled. For example, in the case of a square four-angle column tied together by lacing the approximate radius of gyration, $r = 0.42 h$, where h is the width of the column. For such a column the radius of gyration is directly proportional to the width of the column.

Example 4-12   *Design of Laced Column*

Using two 12-in. miscellaneous channels as shown in Fig. 4-13, design a laced column 26 ft-3 in. long with one end pinned and the other end fixed ($K = 0.80$) to support 320 k. Use A36 steel. Assume adequate welded end connections for lacing.

Solution

Selection of channels. From Table 4-2 the controlling *approximate* radius of gyration $r_x = 0.36h = 0.36(12) = 4.32$ in. and the slenderness ratio

$$\frac{K_x L_x}{r_x} = \frac{0.80(26.25)(12)}{4.32} = 58.3$$

From the AISC *Manual,* Table 3-36, $F_a = 17.59$ ksi. The required area $A = P/F_a = 320/17.59 = 18.19$ in.^{2} and the area for each channel $A = 18.19/2 = 9.10$ in.^{2}

Try two MC12 × 32.9 (for one channel): $A = 9.67$ in.^{2}, $x = 0.867$ in., $r_x = 4.44$ in., $r_y = 0.960$ in. Using the method of Example 4-2, we calculate the radius of gyration with respect to the y axis for the two channels. The transfer distance

$$d_x = \frac{h}{2} - x = 6 - 0.867 = 5.133 \text{ in.}$$

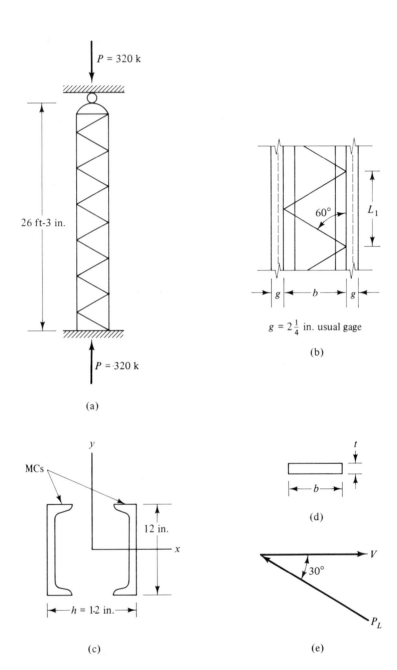

Figure 4-13 Example 4-12.

and from the parallel axis theorem for radiuses of gyration,

$$r_y^2 = d_x^2 + r_{yc}^2 = (5.133)^2 + (0.960)^2$$

$$r_y = 5.22 \text{ in.}$$

Therefore, the controlling slenderness ratio is with respect to the x axis and is equal to 56.8. From the AISC *Manual*, Table 3-36, $F_a = 17.83$ ksi and the allowable load,

$$P = F_a A = 17.73(2)(9.67) = 343 \text{ k} > 320 \text{ k} \qquad \text{OK}$$

Use two MC12 × 32.9.

Local buckling. The channel flanges must satisfy local buckling requirements for an unstiffened element (AISC Specification 1.9.1.2). That is,

$$\frac{b}{t} = \frac{3.500}{0.600} = 5.83 < \frac{95}{\sqrt{F_y}} = \frac{95}{6} = 15.8 \qquad \text{OK}$$

Design of single lacing. From AISC Specification 1.18.2.6 the inclination of the lacing bars must not be less than 60°. In Fig. 4-13(b) the distance between flange holes or the approximate center-to-center weld distance is given by

$$b = h - 2g = 12 - 2(2.25) = 7.5 \text{ in.}$$

The length of channel between lacing connections

$$L_1 = \frac{b}{\sin 60°} = \frac{7.5}{\sin 60°} = 8.7 \text{ in.}$$

For a single MC12 × 32.9 the slenderness ratio

$$\frac{L_1}{r_y} = \frac{8.7}{0.960} = 9.06 < \frac{K_x L_x}{r_x} = 56.8$$

the slenderness ratio for the entire column. According to the AISC Specification 1.18.2.6, the force in the lacing bar

$$V = 0.02P = 0.02(343) = 6.86 \text{ k}$$

or $6.86/2 = 3.43$ on each plane of lacing. The force on one lacing bar to balance the shear force [Fig. 4-13(e)] is given by

$$P_L = \frac{V}{\cos 30°} = \frac{3.43}{\cos 30°} = 3.96 \text{ k}$$

The radius of gyration for the flat lacing bar [Fig. 4-13(d)],

$$r = \sqrt{\frac{I}{A}} = \sqrt{\frac{bt^3/12}{bt}} = 0.289t$$

For single lacing $L_1/r \leq 140$. Solving for the minimum thickness yields

$$t_{min} \geq \frac{L_1}{0.289(140)} = \frac{8.7}{0.289(140)} = 0.215 \text{ in.} \quad (\text{use } \tfrac{1}{4} \text{ in.})$$

With $t = 0.25$ in.,

$$\frac{L_1}{r} = \frac{8.7}{0.289(0.25)} = 120.4$$

From the AISC *Manual*, Table 3-36, $F_a = 10.24$ ksi for this secondary member. Therefore,

$$\text{required } A \geq \frac{P}{F_a} = \frac{3.96}{10.24} = 0.387 \text{ in.}^2$$

$$\text{width} \geq \frac{0.387}{0.25} = 1.547 \text{ in.} \quad (\text{use } 1\tfrac{3}{4} \text{ in.})$$

With welds for the end connections no holes are deducted from the area. Use lacing bar $PL\tfrac{1}{4} \times 1\tfrac{3}{4}$.

Design of end tie plates. From AISC Specification 1.18.2.5 the tie plates should extend along the column a minimum length equal to the distance between lines of fasteners or centers of welds ($b = 7.5$ in.) and the thickness t should be not less than 2 percent of the distance b.

$$t \geq \frac{2}{100} b = \frac{2}{100}(7.5) = 0.15 \text{ in.} \quad (\text{use } \tfrac{3}{16} \text{ in.})$$

Use tie plate $PL\tfrac{3}{16} \times 7\tfrac{1}{2} \times 11\tfrac{1}{2}$.

Example 4-13   *Design of Built-up Column*

Select the angles for a column built up of four corner angles connected or laced together as shown in Fig. 4-14. The member supports a load of 615 k, has a length of 28 ft, and is pinned at the top and bottom. Use A36 steel.

Solution

Selection of angles. The effective length $KL = 1(28)(12) = 336$ in. From Table 4-2 the approximate radius of gyration $r_x = r_y = 0.42h$. Thus

$$\frac{KL}{r} = \frac{KL}{0.42h} = \frac{336}{0.42h} = \frac{800}{h}$$

For various values of h we calculate the slenderness ratio, allowable stress, and required cross-sectional area for each angle. Angles are selected to satisfy the required area. Values are shown in Table (a). For a minimum area, $h = 22$ in. and we try four $L5 \times 5 \times \tfrac{7}{8}$. Because the radius of gyration of the cross section was approximate, this section must be checked.

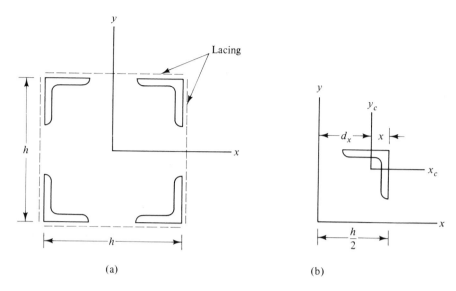

Figure 4-14   Example 4-13.

Properties of L5 × 5 × $\frac{7}{8}$. $A = 7.98$ in.^{2}, $r_x = r_y = 1.49$ in., $x = y = 1.57$ in. Using the method of Example 4-2, we calculate the radius of gyration with respect to the y axis for one angle. In Fig. 4-14(b) we consider one angle. The distance between the two parallel axes y_c and y is

$$d_x = \frac{h}{2} - x = 10 - 1.57 = 8.43 \text{ in.}$$

TABLE (a) for Example 4-13

h (in.)	KL/r	F_a* (ksi)	Req'd $A = \dfrac{615}{4F_a}$ (in.^{2})	Angle Selection
16	50	18.35	8.38	L6 × 6 × $\frac{3}{4}$, $A = 8.44$ in.^{2}
18	44.4	18.83	8.17	L6 × 6 × $\frac{3}{4}$, $A = 8.44$ in.^{2}
20	40	19.19	8.01	L6 × 6 × $\frac{3}{4}$, $A = 8.44$ in.^{2}
22	36.4	19.47	7.90	†L5 × 5 × $\frac{7}{8}$, $A = 7.98$ in.^{2}
24	33.3	19.70	7.80	L5 × 5 × $\frac{7}{8}$, $A = 7.98$ in.^{2}

*From Eq. (4-8) or AISC *Manual,* Table 3-36.
†Design selection.

From the parallel axis theorem

$$r_y^2 = r_{yc}^2 + d_x^2 = (1.49)^2 + (8.43)^2 = 73.28$$

$$r_y = 8.56 \text{ in.}$$

Since the section is symmetrical, $r_x = r_y = 8.56$ in. The slenderness ratio

$$\frac{KL}{r} = \frac{1.0(22)(12)}{8.56} = 30.8$$

From AISC *Manual*, Table 3-36, $F_a = 19.88$ ksi and the allowable load

$$P_a = F_a A = 19.88(4)(7.98) = 635 \text{ k} > 615 \text{ k} \quad \text{OK}$$

Use four L5 × 5 × $\frac{7}{8}$ with $h = 22$ in.

Local buckling. The angle legs must satisfy local buckling requirements for an unstiffened element (AISC Specification 1.9.1.2). That is,

$$\frac{b}{t} = \frac{5}{0.875} = 5.71 < \frac{95}{\sqrt{F_y}} = \frac{95}{6} = 15.5 \quad \text{OK}$$

Design of single lacing. From AISC Specification 1.18.2.6 the inclination of the lacing bars shall be not less than 60°. Using a gage $g = 3$ in. for the angles, the distance between angle holes or the approximate center-to-center weld distance is given by

$$b = h - 2g = 22 - 2(3) = 16 \text{ in.} > 15 \text{ in.}$$

Therefore, we must either use double lacing or single lacing made of angles. Use single lacing made of angles. The length of *column* angles between lacing connections

$$L_1 = \frac{b}{\sin 60°} = \frac{16}{\sin 60°} = 18.5 \text{ in.}$$

For a single L5 × 5 × $\frac{7}{8}$ the slenderness ratio

$$\frac{L_1}{r_z} = \frac{18.5}{0.973} = 19.0 < 30.8$$

the slenderness ratio for the entire column. The shear force in the lacing bars

$$V = 0.02P = 0.02(635) = 12.7 \text{ k}$$

or $12.7/2 = 6.35$ k on each plane of lacing. The force on one lacing bar as shown in Fig. 4-13(e) is given by

$$P_L = \frac{V}{\cos 30°} = \frac{6.35}{\cos 30°} = 7.33 \text{ k}$$

The approximate radius of gyration from Table 4-2 for an angle $r_z = 0.197h$.

For single lacing $L_1/r_z = 140$. Solving for the size of angle h gives

$$h \geq \frac{L_1}{140(0.197)} = \frac{18.5}{140(0.197)} = 0.671 \text{ in.}$$

Try L2 $\times$ 2 $\times$ t, $r_z \approx 0.197(2) = 0.394$ in.

$$\frac{L_1}{r_z} = \frac{18.5}{0.394} = 47.0$$

From the AISC *Manual*, Table 3-36, $F_a = 18.61$ ksi and

$$\text{required } A \geq \frac{P}{F_a} = \frac{7.33}{18.61} = 0.394 \text{ in.}^2$$

Use an L2 $\times$ 2 $\times$ $\frac{1}{8}$, $A = 0.484$ in.^{2}, $r_z = 0.398$ in.

Design of end tie plates

minimum length $b = 16$ in.

The thickness must be not less than 2 percent of distance b:

$$t = \frac{2}{100}(16) = 0.32 \quad (\text{use } \tfrac{3}{8} \text{ in.})$$

Use tie plate PL$\frac{3}{8}$ $\times$ 16 $\times$ 1 ft-9$\frac{1}{2}$ in.

4-12 AXIALLY LOADED COLUMN BASE PLATES

Steel base plates are used under columns to distribute the column load over a sufficient area of the supporting footing so that the footing is not overstressed. The design method from the AISC *Manual* assumes that the column load is distributed uniformly over a rectangle $0.95d$ by $0.80h$ [Fig. 4-15(a)]. The base plate then distributes this load uniformly to the footing under the plate as shown in Fig. 4-15(b). The plate has a tendency to bend upward around the outside edges. The thickness of the base plate is determined by the resulting bending stress. When n and m are not small, the critical sections for bending moments occur at a distance n from the edge of the plate in one direction and at a distance m in the other direction. The bending moments on cantilever beams of unit width and spans m or n [Fig. 4-15(c)] are given by

$$M = \frac{f_p n^2}{2} \quad \text{and} \quad M = \frac{f_p m^2}{2}$$

The bending stresses for the two directions are

$$f_b = \frac{Mc}{I} = \frac{\dfrac{f_p n^2}{2}\left(\dfrac{t}{2}\right)}{\dfrac{1}{12}(1)t^3} = \frac{3f_p n^2}{t^2} \tag{4-11}$$

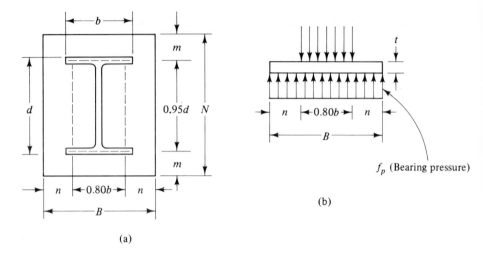

(a)

(b)

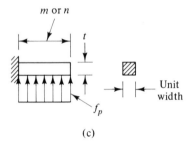

(c)

Figure 4-15   Column base plate.

and

$$f_b = \frac{3f_p m^2}{t^2}$$ (4-12)

AISC Specification 1.5.1.4.3 permits an allowable stress of $0.75F_y$ for bending of a rectangular cross section. To find the required plate thickness, we equate the allowable bending stress to the bending stress given by Eqs. (4-11) and (4-12). Solving for the plate thicknesses yields

$$t = \sqrt{\frac{3f_c n^2}{0.75F_y}}$$ (4-13)

and

$$t = \sqrt{\frac{3f_c m^2}{0.75F_y}}$$ (4-14)

When n and m are small, the highest stress in the base plate will occur at the face of the column web halfway between the flanges. For this case, the thickness of the base plate can be calculated from the bending of a flat plate with edge conditions that approximate those of the part of a base plate between the flanges of a column. The AISC *Manual* provides a simple method for making this calculation. An equivalent cantilever dimension n' is tabulated for each column section. The thickness of the base plate is then determined from either Eq. (4-13) or (4-14) with n or m replaced by n'. Column base plate design is outlined as follows:

1. Determine the allowable bearing stress F_p for the footings according to AISC Specification 1.5.5.
2. Determine the required plate area.

$$\text{required } A \geq \frac{P}{F_p}$$

where A = required plate area, that is, $A = BN$
P = column load
F_p = allowable bearing stress

3. Select B and N, preferably rounded to the nearest inch such that n and m are nearly equal.
4. Determine the actual bearing pressure

$$f_p = \frac{P}{BN}$$

5. Determine n and m from Fig. 4-20(a). Thus

$$n = \frac{B - 0.80b}{2} \qquad m = \frac{N - 0.95d}{2}$$

and n' is determined from Table C, p. 3-100, of the AISC *Manual.*

6. Use the largest value of n, m, or n' to determine the thickness of the plate.

$$t = (n, m, \text{ or } n') \sqrt{\frac{3f_c}{0.75F_y}}$$

Example 4-14

A W10 × 60 column of A36 steel supports an axial load of 320 k. Design a base plate for the column if the supporting concrete has an ultimate strength $f'_c = 3.0$ ksi. The concrete support has a cross-sectional area much larger than the bearing plate area.

Solution.   From AISC Specification 1.5.5,

$$F_p = 0.35f'_c \sqrt{\frac{A_1}{A_2}} \leq 0.7f'_c$$

Since $A_1 \gg A_2$,

$$F_p = 0.7f'_c = 0.7(3.0) = 2.1 \text{ ksi}$$

The required area

$$\text{required } A \geq \frac{P}{F_p} = \frac{320}{2.1} = 152.4 \text{ in.}^2$$

Select $N = 13$ in. and $B = 12$ in. The area

$$A = NB = 13(12) = 156 \text{ in.}^2$$

Then

$$m = \frac{N - 0.95d}{2} = \frac{13 - 0.95(10.22)}{2} = 1.646 \text{ in.}$$

and

$$n = \frac{B - 0.80b}{2} = \frac{12 - 0.80(10.08)}{2} = 1.968 \text{ in.}$$

and from Table C, p. 3-100, of the AISC *Manual*

$$n' = 3.92 \text{ in.}$$

The actual bearing pressure

$$f_p = \frac{P}{NB} = \frac{320}{13(12)} = 2.05 \text{ ksi}$$

and the plate thickness

$$t \geq n' \sqrt{\frac{3f_p}{0.75F_y}} = 3.92 \sqrt{\frac{3(2.05)}{0.75(36)}} = 1.87 \text{ in.} \quad \left(\text{use } 1\tfrac{7}{8} \text{ in.}\right)$$

Use a $\text{PL}1\tfrac{7}{8} \times 12 \times 1$ ft-1 in.

PROBLEMS

Use the AISC *Specifications* for the following problems. Where end conditions for columns are given, use the recommended effective length. In design problems determine the lightest section.

4-1. Determine the allowable axial load for a compression member made of A36 steel with cross section and effective length as follows:
   (a) W12 × 65, $KL = 22$ ft.
   (b) S6 × 12.5, $KL = 4$ ft 6 in.
   (c) WT9 × 43, $KL = 11$ ft 6 in.
   (d) M6 × 20, $KL = 19$ ft 6 in.

4-2. Repeat Prob. 4-1 using A441 steel with $F_y = 50$ ksi.

4-3. Determine the allowable axial compressive load for a column made of A36 steel with a built-up section, end conditions, and length as shown.

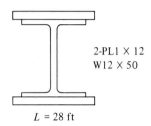

2-PL1 × 12
W12 × 50

$L = 28$ ft

(a) Fixed ends

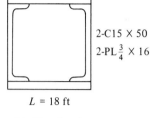

2-C15 × 50
2-PL$\frac{3}{4}$ × 16

$L = 18$ ft

(b) Pinned ends

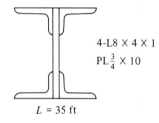

4-L8 × 4 × 1
PL$\frac{3}{4}$ × 10

$L = 35$ ft

(c) Fixed at bottom,
    pinned at top

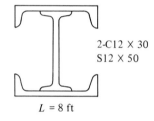

2-C12 × 30
S12 × 50

$L = 8$ ft

(d) Fixed ends

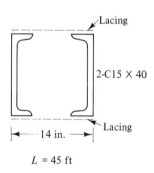

2-C15 × 40

14 in.

$L = 45$ ft

(e) Pinned ends

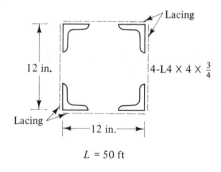

12 in.

4-L4 × 4 × $\frac{3}{4}$

12 in.

$L = 50$ ft

(f) Pinned ends

Prob. 4-3

4-4. Repeat Prob. 4-3 using A572 Grade 50 steel.

4-5. A pin-ended column is supported at midheight perpendicular to the y axis of its cross section. Determine the allowable axial load if the member is made of A36 steel and has a cross section and length as follows:

(a) W14 × 68, $L = 18$ ft 3 in.
(b) S6 × 17.25, $L = 9$ ft 5 in.
(c) M6 × 20, $L = 8$ ft 4 in.
(d) WT8 × 50, $L = 20$ ft.

4-6. Repeat Prob. 4-5 using A572 Grade 42 steel.

4-7. A compression member in a truss is made of A36 steel and is formed by two angles with long legs back to back as shown. Determine the allowable axial load

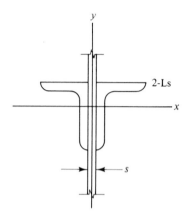

Prob. 4-7

that can be supported by the member if the ends are pinned and it is braced at midlength perpendicular to the y axis. Angles, angle separation, and length are as follows:

(a) Two L4 × $3\frac{1}{2}$ × $\frac{1}{2}$, $s = \frac{3}{8}$ in., $L = 12$ ft.
(b) Two L6 × 4 × $\frac{5}{8}$, $s = \frac{3}{8}$ in., $L = 20$ ft.
(c) Two L7 × 4 × $\frac{3}{4}$, $s = \frac{3}{4}$ in., $L = 18$ ft.
(d) Two L8 × 4 × $\frac{3}{4}$, $s = 1$ in., $L = 32$ ft.

4-8. Repeat Prob. 4-7 using A572 Grade 50 steel.

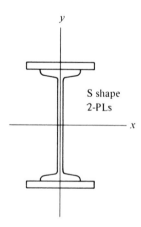

S shape
2-PLs

Prob. 4-9

4-9. A column made of A36 steel is formed by an S shape and two plates as shown. Determine the allowable axial load the column can support if the ends are pinned and it is braced at midheight perpendicular to the y axis. Plates, S shapes, and length are as follows:

(a) Two $PL\frac{3}{4} \times 8$, $S8 \times 23$, $L = 24$ ft.
(b) Two $PL\frac{3}{4} \times 10$, $S10 \times 35$, $L = 26$ ft.
(c) Two $PL\frac{3}{4} \times 12$, $S12 \times 50$, $L = 27$ ft.
(d) Two $PL\frac{3}{4} \times 15$, $S15 \times 50$, $L = 28$ ft.

4-10. Repeat Prob. 4-9 using A572 Grade 60 steel.

4-11. A column made of A36 steel is formed by four angles laced together as shown.

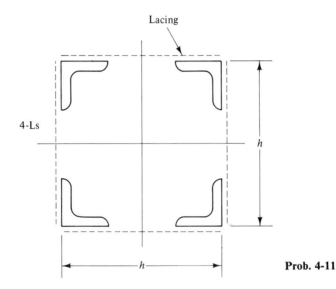

Lacing

4-Ls

h

h

Prob. 4-11

Determine the allowable axial load the column can support. Angles, depth of section, and effective length are as follows:

(a) Four $L4 \times 4 \times \frac{1}{2}$, $h = 16$ in., $KL = 25$ ft.
(b) Four $L4 \times 4 \times \frac{5}{8}$, $h = 18$ in., $KL = 28$ ft.
(c) Four $L5 \times 5 \times \frac{3}{4}$, $h = 20$ in., $KL = 30$ ft.
(d) Four $L6 \times 6 \times \frac{3}{4}$, $h = 22$ in., $KL = 32$ ft.

4-12. Repeat Prob. 4-11 using A572 Grade 50 steel.

4-13. Select the lightest W14 column section made of A36 steel with pinned ends to support a load with a length as follows:

(a) $P = 400$ k, $L = 25$ ft.
(b) $P = 450$ k, $L = 27$ ft.
(c) $P = 500$ k, $L = 29$ ft.
(d) $P = 550$ k, $L = 31$ ft.

4-14. Repeat Prob. 4-13 using A588 steel with $F_y = 50$ ksi.

4-15. Select the lightest W12 column section to support a load of 300 k. The member is 35 ft long and is fixed at the bottom and pinned at the top. Use A36 steel.

4-16. Repeat Prob. 4-15 using A572 Grade 50 steel.

4-17. Select the most economical W12 column section made of A36 steel with pinned ends if the column is braced at midheight perpendicular to the y axis of the cross section. The column load and length are as follows:
(a) $P = 400$ k, $L = 20$ ft.
(b) $P = 450$ k, $L = 22$ ft.
(c) $P = 500$ k, $L = 24$ ft.
(d) $P = 550$ k, $L = 26$ ft.

4-18. Repeat Prob. 4-17 using A572 Grade 50 steel.

4-19. Select the lightest double-angle compression member made of A36 steel for the top chord of a truss (see Prob. 4-7). In the plane of the truss, cross members provide bracing at intervals of length L. For out-of-plane buckling, bracing is provided at intervals of $2L$. The member is subject to wind loads that are included in the design load. The design load, angle separation, and length L are as follows:
(a) $P = 66$ k, $s = \frac{3}{8}$ in., $L = 5$ ft.
(b) $P = 120$ k, $s = \frac{3}{8}$ in., $L = 5$ ft.
(c) $P = 140$ k, $s = \frac{3}{8}$ in., $L = 6$ ft.
(d) $P = 200$ k, $s = \frac{3}{8}$ in., $L = 6$ ft.

4-20. Repeat Prob. 4-19 using A572 Grade 50 steel.

4-21. Select the lightest structural tee (WT) compression member made of A36 steel for the top chord of a truss. The bracing for in-plane and out-of-plane buckling is provided at intervals of length L. The member is subject to wind loads that are included in the design load. The design load and length L are as follows:
(a) $P = 120$ k, $L = 8$ ft.
(b) $P = 85$ k, $L = 10$ ft.
(c) $P = 125$ k, $L = 9$ ft.
(d) $P = 45$ k, $L = 15$ ft.

4-22. Repeat Prob. 4-21 using A572 Grade 50 steel.

4-23. Design a column formed by four corner angles connected together as shown in Prob. 4-11. The member supports a load of 520 k, has a length of 17 ft-2 in., and is fixed at the bottom and free at the top. Use A36 steel.

4-24. Repeat Prob. 4-23 using A572 Grade 42 steel.

4-25. Design a column built up of two 10-in. channels laced together as shown to

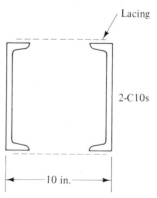

Lacing

2-C10s

10 in.

Prob. 4-25

support a load of 180 k. The member is 35 ft long and is fixed at the bottom and pinned at the top. Use A572 Grade 50 steel.

4-26. Repeat Prob. 4-25 using A572 Grade 60 steel.

4-27. Four $5 \times 5 \times \frac{1}{2}$ angles made of A36 steel are used to form a column section as shown in Prob. 4-11. The member has a length of 20 ft and a depth h of 20 in.
 (a) Determine the allowable axial load for the column.
 (b) Design single lacing and tie plates assuming that connections are made to the angle with $\frac{3}{4}$-in.-diameter bolts.

4-28. Repeat Prob. 4-27 using A572 Grade 50 steel.

4-29. Two C10 × 50 channels separated by 12 in. are used to form a column section as shown in Prob. 4-25. The member is made of A36 steel, has a length of 30 ft, and is fixed at the bottom and pinned at the top.
 (a) Determine the allowable load.
 (b) Design single lacing and tie plates assuming connections are made to the channel with $\frac{3}{4}$-in.-diameter bolts.

4-30. Repeat Prob. 4-29 using A572 Grade 50 steel.

4-31. A column is supported by a base plate that rests on a reinforced concrete footing. The base plate is made of A36 steel. What is the maximum allowable load the plate can support if the column section, base plate size, and allowable bearing pressure are as follows?
 (a) W14 × 74, PL2 × 22 × 2 ft-3 in., $F_p = 0.750$ ksi.
 (b) W14 × 90, PL1$\frac{7}{8}$ × 19 × 2 ft-0 in., $F_p = 1.125$ ksi.
 (c) W12 × 58, PL1$\frac{1}{2}$ × 16 × 1 ft-8 in., $F_p = 1.125$ ksi.
 (d) W12 × 65, PL1$\frac{7}{8}$ × 22 × 2 ft-0 in., $F_p = 0.750$ ksi.

4-32. Repeat Prob. 4-31 using A572 Grade 50 steel and allowable bearing pressures for parts (a) and (d) of 1.000 ksi and parts (b) and (c) of 1.600 ksi.

4-33. A column supporting an axial load has a base plate that rests on a concrete footing. Design a base plate of A36 steel for the column if the column section, axial load, and allowable bearing pressure on the concrete are as follows:
 (a) W8 × 58, $P = 250$ k, $F_p = 1.125$ ksi.
 (b) W10 × 88, $P = 430$ k, $F_p = 1.500$ ksi.
 (c) W12 × 96, $P = 500$ k, $F_p = 1.250$ ksi.
 (d) W14 × 109, $P = 590$ k, $F_p = 1.750$ ksi.

4-34. Repeat Prob. 4-33 using A572 Grade 50 steel.

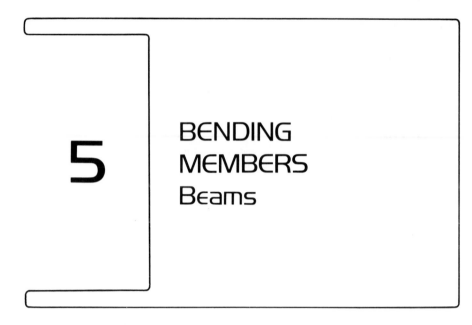

5 BENDING MEMBERS
Beams

Bending members are commonly called *beams*. They support loads that are mainly applied perpendicular or transverse to the axis of the member. The loads produce bending moments and shear forces in the beam.

Beams in structures may also be referred to by names that suggest their function in the structure. Some of the names in common use are as follows:

Girders: usually indicate a major beam frequently at wide spacing that supports smaller beams

Joists: closely spaced beams supporting the floors and roofs of buildings, often with truss-type webs

Purlins: roof beams usually supported by trusses

Rafters: roof beams usually supported by purlins

Lintels: beams over window or door openings that support the wall above

Girts: horizontal wall beams used to support wall coverings on the side of an industrial building

Spandrel beams: beams around the outside perimeter of a floor that support the exterior walls and the outside edge of the floor

The beam-column, a member that supports both transverse and axial loads, is considered in Chapter 6.

5-2 TYPES OF FAILURE IN BEAMS

The transverse loads on beams cause bending moments and shear forces. Failure of beams can occur in several ways. Bending failure begins with excessive deflection followed by collapse due to crushing of the compression flange or fracture of the tension flange of the beam. Instead of failure due to crushing, the compression flange may fail by a column-like action with sidesway or lateral buckling. Collapse would probably follow the lateral buckling. Shear failure would most likely be observed as buckling of the web of the beam near locations of high shear forces. Near reactions or concentrated loads the beam can fail locally due to crushing or buckling of the web. Large beam deflections can also represent failure when the intended use of the beam places limits on deflections.

Beam design usually requires the selection of a lightest member that is safe from the various types of failures. The methods required to achieve this objective are discussed in the following sections.

5-3 ELASTIC BENDING OF BEAMS

The bending stress for a beam is given by the bending stress or flexure formula

$$f_b = \frac{Mc}{I} \qquad (5\text{-}1)$$

where f_b = maximum bending stress, ksi
  M = bending moment at the section in question, k-in.
  c = distance from the centroidal (neutral) axis of the beam to the extreme top or bottom fiber of the cross section, in.
  I = moment of inertia of the cross-sectional area of the beam about the centroidal axis, in.[4]

The derivation of the bending stress or flexure formula can be found in any elementary strength-of-materials textbook.

For a given cross section both I and c are constants. We define a new constant, $I/c = S$, called the *elastic section modulus*. Equation (5-1) can be rewritten in terms of the elastic section modulus as follows:

$$f_b = \frac{M}{I/c} = \frac{M}{S} \qquad (5\text{-}2)$$

where f_b = maximum bending stress, ksi
  M = bending moment at the section in question, k-in.
  S = section modulus of the cross-sectional area of the beam about the centroidal axis, in.[3]

Example 5-1   *Bending Stress for W Section Beam*

Determine the maximum bending stress f_b in a W14 × 82 beam if the maximum bending moment is 204 k-ft. Use Eq. (5-1).

Solution.   Properties of W14 × 82: $d = 14.31$ in., $I_x = 882$ in.[4]. The neutral axis is at the middle of the cross section; therefore, $c = d/2 = 14.31/2 = 7.155$ in. The maximum bending moment is $M = 204$ k-ft $= 204(12) = 2448$ k-in. and the maximum bending stress

$$f_b = \frac{Mc}{I} = \frac{2448(7.155)}{882} = 19.9 \text{ ksi}$$

Example 5-2   *Bending Stress for W Section Beam*

Determine the maximum bending stress f_b in a W16 × 100 beam if the maximum bending moment is 280 k-ft. Use Eq. (5-2).

Solution.   Properties of W16 × 100: $S_x = 175$ in.[3]. The maximum bending stress

$$f_b = \frac{M}{S} = \frac{280(12)}{175} = 19.2 \text{ ksi}$$

5-4 INELASTIC BEHAVIOR OF STEEL BEAMS

Elastic or allowable stress design is emphasized in this chapter. However, a brief introduction to inelastic behavior of steel beams is essential for understanding provisions in the AISC *Specifications* that permit different allowable stresses for the various types of cross sections and bending axes of beams.

An increasing bending moment is applied to a wide-flange steel beam. Various stages of the resulting stress distribution at a cross section of the beam are shown in Fig. 5-1. The steel in the beam is assumed to have the idealized stress–strain behavior of Fig. 5-2. This idealization is generally considered valid for steels with yield stresses up to 65 ksi.

When the bending moment is first applied, the bending stresses are proportional to the distance from the neutral axis [Fig. 5-1(a)]. The stress on the beam cross section is entirely elastic. With continued loading the extreme fibers yield and the applied moment is equal to the *yield moment M_y* [Fig. 5-1(b)]. The yield moment can be calculated from Eq. (5-2) by equating the yield stress to the maximum bending stress. That is,

$$F_y = \frac{M_y}{S} \qquad \text{or} \qquad M_y = F_y S \tag{5-3}$$

where S is the elastic section modulus with respect to the bending axis.

With further increases in the load the stresses on the outer fibers remain constant at the yield stress F_y and the stresses on the inner fibers are still

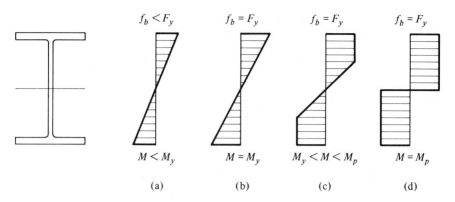

Figure 5-1   Stress distributions for a beam subject to an increasing bending moment.

proportional to the distance from the neutral axis [Fig. 5-1(c)]. The outer fibers are now plastic while the inner fibers are still elastic.

Continued loading will cause all the fibers to yield [Fig. 5-1(d)]. The entire cross section is plastic and the applied moment is equal to the *plastic moment* M_p. The plastic moment can be determined from an equation

$$M_p = F_y \sum yA = F_y Z \tag{5-4}$$

where Z is the plastic section modulus with respect to the bending axis.

The plastic section modulus for various structural shapes is given in the AISC *Manual.* The *Manual* also has formulas for the plastic section modulus of various geometric shapes (pp. 6-19 through 6-22).

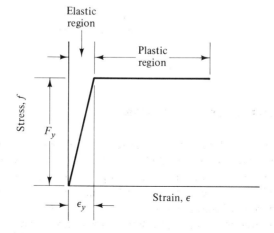

Figure 5-2   Idealized stress–strain diagram for steel.

Shape factor.   The ratio of M_p to M_y depends only on the cross-sectional properties of a member and is called the *shape factor f.* That is,

$$f = \frac{M_p}{M_y} = \frac{Z}{S} \tag{5-5}$$

Shape factors for W sections vary between 1.10 and 1.18, with the most usual value about 1.12. For round sections the shape factor is 1.70 and for rectangles 1.50.

Example 5-3   *Shape Factor for a W Shape*

Determine the shape factor for the W18 × 119 with respect to
  (a) The x axis.
  (b) The y axis.

Solution.   **(a)** Properties from the AISC *Manual*: $Z_x = 261$ in.³, $S_x = 231$ in.³. Therefore, the shape factor

$$f = \frac{Z_x}{S_x} = \frac{261}{231} = 1.13$$

  (b) Properties from the AISC *Manual*: $Z_y = 69.1$ in.³, $S_y = 44.9$ in.³. The shape factor

$$f = \frac{Z_y}{S_y} = \frac{69.1}{44.9} = 1.54$$

For bending about the y axis the W section consists essentially of three rectangles (Fig. 5-3). Thus the shape factor is approximately the same as for a rectangle ($f = 1.5$).

Figure 5-3   Example 5-3.

5-5 LATERAL SUPPORTS FOR BEAMS

Loads on a beam cause bending and the resulting tension in one flange and compression in the other flange of the beam. The compression flange may fail by a column-like action with lateral buckling. To prevent buckling, lateral supports or bracing are necessary. Most beams have some lateral support. However, the degree of lateral support is often a matter of engineering judgment. If the compression flange of a beam supporting a floor is encased or connected by shear connectors to the floor (Fig. 5-4), the beam is without

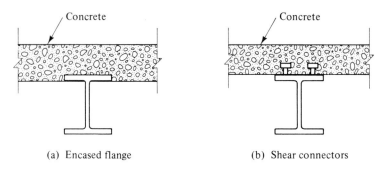

(a) Encased flange   (b) Shear connectors

Figure 5-4   Laterally supported beams.

question laterally supported. Cross beams framed into the sides of the beams at intervals provide lateral support if the cross beams are adequately braced as shown in Fig. 5-5. Steel roof decks attached by welds or screws at close spacing provide good lateral support for the top flanges of the supporting members.

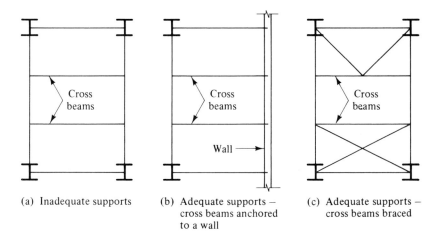

(a) Inadequate supports   (b) Adequate supports –   (c) Adequate supports –
                                   cross beams anchored   cross beams braced
                                   to a wall

Figure 5-5   Beams with lateral supports.

   In other cases the degree of lateral support may be unclear. If light-gage steel decking is spot welded to the supporting beams, partial lateral support can probably be assumed. On the other hand, if the decking rests but is not attached to the supporting beams, partial lateral support due to friction can be assumed only if the loads remain fixed in location and there is no chance of vibration.
   Each case encountered requires careful consideration. In doubtful cases it is conservative and safe to assume no lateral supports.

5-6 DESIGN OF LATERALLY SUPPORTED BEAMS

The basic allowable stress in tension and compression is $0.6F_y$. For sections that can develop their full plastic moments the allowable stress has been increased by 10 percent, to $0.66F_y$. This increase is for sections with a plastic moment 10 to 18 percent greater than the yield moment (Sec. 5-4).

AISC Specification 1.5.1.4.1 gives the allowable stress on the extreme fibers of hot-rolled sections that are symmetric about and loaded in the plane of their minor axis (y axis) as

$$F_b = 0.66F_y$$

provided that the section is *compact* and has adequate *lateral bracing*.

Compact section. To qualify as compact, the section must develop its full plastic moment before localized buckling can occur. A section is compact if it satisfies the following requirements (AISC Specification 1.5.1.4.1):

1. The width-to-thickness ratio of unstiffened projecting elements of the compression flange must not exceed $65/\sqrt{F_y}$. That is,

$$\frac{b_f}{2t_f} \le \frac{65}{\sqrt{F_y}}$$

Solving for F_y and setting $F_y = F_y'$ in this expression, we have

$$F_y' = \left(\frac{130t_f}{b_f}\right)^2 \quad \text{ksi}$$

where F_y' is the yield stress for which the flanges of a given section will be compact. Values of F_y' are given with design properties and the allowable stress design selection table for beams in the AISC *Manual.*

2. The depth-to-thickness ratio of web must not exceed $640/\sqrt{F_y}$. That is,

$$\frac{d}{t_w} \le \frac{640}{\sqrt{F_y}}$$

Solving for F_y and setting $F_y = F_y''$ in this expression, we have

$$F_y'' = \left(\frac{640t_w}{d}\right)^2 \quad \text{ksi}$$

where F_y'' is the yield stress for which the web of a given section will be compact.

All current rolled shapes with yield stresses up to 65 ksi have compact webs ($F_y'' = 65$ ksi). Values are not shown in the AISC *Manual* because plastic behavior is not recognized by the AISC for steels with $F_y > 65$ ksi.

Adequate bracing.   To be adequately braced, the laterally unsupported length L_b of the compression flange of the member must be the smaller of the following distances:

$$L = \frac{76 b_f}{12\sqrt{F_y}} \quad \text{ft} \qquad \text{or} \qquad L = \frac{20,000}{12(d/A_f)F_y} \quad \text{ft}$$

The smaller of these distances is called L_c. Values of L_c are given in the allowable stress design selection tables for beams of the AISC *Manual.*

The bending stress permitted by the AISC *Specifications* for a noncompact section is $0.6F_y$. However, if $b_f/2t_f$ exceeds $65/\sqrt{F_y}$ but is less than $95/\sqrt{F_y}$, AISC Specification 1.5.1.4.2 provides a linear transition for the bending stress between $0.66F_y$ and $0.6F_y$. The transition bending stresses are given by the equation

$$F_b = F_y\left[0.79 - 0.002\left(\frac{b_f}{2t_f}\right)\sqrt{F_y}\right]$$

The equation does not apply to hybrid girders and members of A514 steel.

The maximum moment may be found by drawing the shear and moment diagrams, using the method of sections, or from formulas. Each of the methods is illustrated in the following examples.

Example 5-4   *Simply Supported Beam with Concentrated Load*

A simply supported beam with a span of 15 ft has a 6-k load 5 ft from one end. Select a beam section of A36 steel that can safely support the load. Assume full lateral support for the compression flange.

Solution.   Neglect the weight of the beam. From Beam Diagrams and Formulas, case 8, of the AISC *Manual,* the maximum moment is under the concentrated load and given by

$$M_{max} = \frac{Pab}{L} = \frac{6(5)(10)}{15} = 20 \text{ k-ft}$$

For A36 steel the allowable stress for a compact section with full lateral support is

$$F_b = 0.66F_y = 24 \text{ ksi}$$

Note that a rounded value for F_b from Appendix A, Table 1 of the AISC *Specifications* is used. Thus the required section modulus

$$\text{required } S = \frac{M}{F_b} = \frac{20(12)}{24} = 10 \text{ in.}^3$$

From the allowable stress design selection table of the AISC *Manual,* the lightest beam with a section modulus larger than 10 in.^{3} is the M12 × 11.8 with a section modulus of 12.0 in.^{3}. The section is compact, $F_y' > 36$ ksi.

The weight of the beam would add an additional moment under the load. From the AISC *Manual* at a distance along the beam of $x = 5$ ft,

$$M_x = \frac{wx}{2}(L - x) = \frac{0.0118(5)(15 - 5)}{2} = 0.295 \text{ k-ft}$$

The maximum moment including the weight of the beam $M_{max} = 20.0 + 0.295 = 20.30$ k-ft and the required section modulus $S = 20.30(12)/24 = 10.15 \text{ in.}^3 < 12 \text{ in.}^3$. Hence use an M12 × 11.8.

In this case the weight of the beam added 1.5 percent to the required section modulus. Unless the required section modulus is close to the section modulus of the selected section, it is safe to neglect the weight of the beam. However, beam weights should be included in the design of members that support the beam.

Example 5-5   *Simply Supported Beam with Uniform Load*

A simply supported beam with a span of 20 ft has a uniform load of 6 k/ft over the entire span. Select a beam that can safely support the load. Assume full lateral support.

(a) Use A36 steel.
(b) Use A572 Grade 42 steel.

Solution.   Neglect the weight of the beam. The maximum moment from the AISC *Manual* is at the middle of the beam and given by

$$M_{max} = \frac{wL^2}{8} = \frac{6(20)^2}{8} = 300 \text{ k-ft}$$

(a) For A36 steel the allowable bending stress for a compact section is $F_b = 0.66F_y = 24$ ksi and the required section modulus

$$S = \frac{M}{F_b} = \frac{300(12)}{24} = 150 \text{ in.}^3$$

From the design selection table the lightest beam with a section modulus larger than 150 in.^{3} is the W24 × 68 with a section modulus of 154 in.^{3}. The section is compact, $F'_y > 36$ ksi. Checking the beam by including the weight, the maximum moment $M = wL^2/8 = 6.068(20)^2/8 = 303.4$ k-ft. The required section modulus $S = M/F_b = 303.4(12)/24 = 153.4 \text{ in.}^3 < 154 \text{ in.}^3$. Therefore, use a W24 × 68.

(b) For A572 Grade 42 steel the allowable bending stress for a compact section is $F_b = 0.66F_y = 0.66(42) = 27.7$ ksi. The required section modulus

$$S = \frac{M}{F_b} = \frac{300(12)}{27.7} = 130 \text{ in.}^3$$

From the design selection table the lightest beam with a section modulus larger than 130 in.^{3} is the W24 × 62, with a section modulus $S = 131 \text{ in.}^3$. The maximum moment including the weight of the beam $M = wL^2/8 =$ 6.062(20)^2/8 = 303.1 k-ft and the required section modulus is $S =$ 303.1(12)/27.7 = 131.3 in.^{3} > 131 in.^{3}. Use of the W24 × 62 would result in an overstress of only 0.25 percent and is probably acceptable. Use a W24 × 62 or if not acceptable, a W21 × 68.

Example 5-6 *Simply Supported Beam with Concentrated Load and Partial Uniform Load*

A simply supported beam has a span and loads as shown in Fig. 5-6(a). Select a beam section of A572 Grade 50 steel that can safely support the loads. The beam has full lateral support.

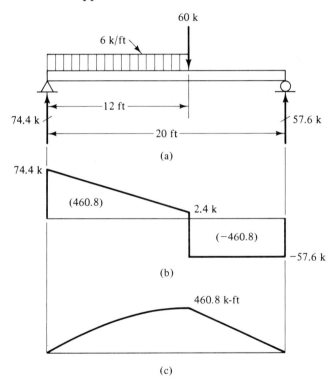

Figure 5-6   Example 5-6.

Solution.   The reactions were calculated and are displayed in Fig. 5-6(a). The shear diagram is shown in Fig. 5-6(b) and the moment diagram in Fig. 5-6(c). From the moment diagram the maximum moment

$$M_{max} = 460.8 \text{ k-ft}$$

For A572 Grade 50 steel the allowable stress for a compact section is

$F_b = 0.66F_y = 0.66(50) = 33$ ksi and the required section modulus

$$S = \frac{M}{F_b} = \frac{460.8(12)}{33} = 167.6 \text{ in.}^3$$

From the design selection table the lightest section is a W24 × 76 with a section modulus of 176 in.³. The section is 5.0 percent larger than required. Therefore, the beam need not be checked for weight. The section is compact, $F_y' > 50$ ksi. Use a W24 × 76.

Example 5-7   *Beam with Overhang*

A beam with an overhanging end has loads and span as shown in Fig. 5-7(a). Select a W section of A36 steel that can safely support the loads. Assume full lateral support of the compression flange.

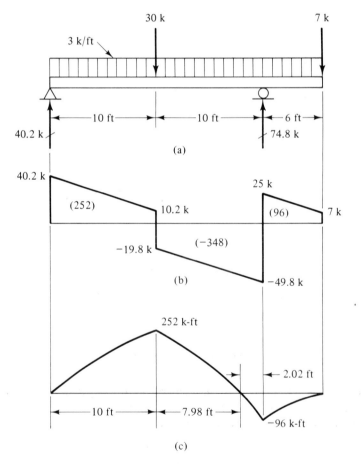

Figure 5-7   Example 5-7.

Solution. The reactions were calculated and are displayed in Fig. 5-7(a). The shear diagram is shown in Fig. 5-7(b) and the moment diagram in Fig. 5-7(c). From the moment diagram the maximum moment

$$M_{\max} = 252 \text{ k-ft}$$

For A36 steel the allowable stress for a compact section is 24 ksi and the required section modulus is $S = M/F_b = 252(12)/24 = 126 \text{ in.}^3$. From the design selection table the lightest beam is a W24 × 62 with a section modulus of 131 in.^{3}. The beam must be checked. The additional moment at a distance $x = 10$ ft from the left end of the beam, due to the weight, from the AISC *Manual* is given by

$$M_x = \frac{wx}{2L}(L^2 - a^2 - xL)$$

$$M_{10} = \frac{0.062(10)}{2(20)}[(20)^2 - (6)^2 - 10(20)] = 2.54 \text{ k-ft}$$

Hence the maximum moment including the weight of the beam $M_{\max} = 252 + 2.54 = 254.54$ k-ft. The required section modulus $S = 254.54(12)/24 = 127.3 \text{ in.}^3 < 131 \text{ in.}^3$. Therefore, use a W24 × 62.

5-7 SHEAR STRESS IN BEAMS

Transverse forces on beams cause shear stresses that can be determined from the formula

$$f_v = \frac{QV}{Ib} \tag{5-6}$$

where F_v = longitudinal shear stress, ksi
  V = shear force at a cross section, k
  $Q = \sum Ay$ = statical moment about the neutral axis of the partial area of the cross section lying either above or below the line where the shear stress is being determined, in.^{3}
  I = moment of inertia about the neutral axis of the cross section, in.^{4}
  b = width of the section where the shear stress is being determined, in.

The derivation of the shear stress formula can be found in any elementary strength-of-materials textbook.

For rectangular beams, the maximum shearing stress occurs at the neutral axis and is given by

$$f_v = \frac{3V}{2A} \tag{5-7}$$

where $A = bh$ = area of the cross section.

For rolled and fabricated shapes, AISC Specification 1.5.1.2.1 permits the shearing stress to be determined by the formula

$$f_v = \frac{V}{dt_w} \tag{5-8}$$

where d = depth of the section, in.
    t_w = thickness of the web, in.
and specifies an allowable shearing stress

$$F_v = 0.4F_y$$

Example 5-8 *Comparison of Maximum and Average Shearing Stress in Rolled Beams*

If an external shear force $V = 100$ k acts on a W14 × 120 beam, calculate the following:
    (a) The maximum shear stress from Eq. (5-6).
    (b) The average shear stress from Eq. (5-8).
    (c) The percentage difference between the maximum shear stress and the average shear stress.

Solution. **(a)** Properties of the W14 × 120 are from the AISC *Manual.* The cross section of the beam is shown in Fig. 5-8. The area above the neutral axis was divided into two rectangles and the centroid for each rectangle was determined. Areas and centroids are illustrated in the figure. The statical moment

$$Q = \sum Ay = A_1y_1 + A_2y_2$$
$$= 14.67(0.940)(6.77) + 6.30(0.590)(3.15)$$
$$= 105.1 \text{ in.}^3$$

The maximum shear stress

$$f_{v\,max} = \frac{VQ}{Ib} = \frac{100(105.1)}{1380(0.590)} = 12.90 \text{ ksi}$$

(b) The average shear stress

$$f_v = \frac{V}{dt_w} = \frac{100}{14.48(0.590)} = 11.71 \text{ ksi}$$

(c) The percentage difference is given by

$$\frac{100(12.90 - 11.71)}{12.9} = 9.2\%$$

The difference between the maximum shear stress and average shear stress for rolled steel sections range in value from approximately 10 to 15 percent.

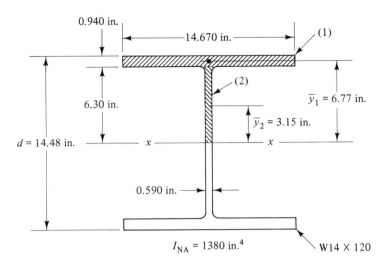

$I_{NA} = 1380$ in.^{4} W14 × 120

Figure 5-8   Example 5-8.

Example 5-9   *Check of Shear Stress in Beams*

Check the shear stress for the beam of Example 5-7.

Solution.   Properties of the W24 × 62: $d = 23.74$ in., $t_w = 0.430$ in.   The maximum shear force from Fig. 5-7(b), $V_{max} = 49.8$ k.   From Eq. (5-8),

$$f_v = \frac{V}{dt_w} = \frac{49.8}{23.74(0.430)}$$

$$= 4.88 \text{ ksi} < F_v = 0.4F_y = 0.4(36) = 14.4 \text{ ksi}   \text{OK}$$

Hence the W24 × 62 beam is satisfactory in shear.

In most cases beams are selected on the basis of bending stress. The shear stress is then checked and found to be less than the allowable value. Some exceptions are beams with heavy loads near beam supports and beams that have been notched or coped. In cases where the shear stress does control, a beam with a large enough web area is selected so that the shear stress is less than the allowable value.

5-8 HOLES IN BEAMS

For tension members the area of holes is deducted to find the net area as discussed in Chapter 3. For columns, fasteners fill most of the space in the holes and transmit the compressive load across the holes. No deduction is made for hole areas in columns.

Any type of hole in a beam will weaken the beam and should be avoided. If that is not possible, the holes should be placed in the web where the bending moment is large or in the flange where the shear force is large. However, tests seem to indicate that unless substantial reductions are made in the cross-sectional area, no reduction need be made in the bending capacity of a beam.

Flange holes.   AISC Specification 1.10.1 provides that no reduction for fastener holes in either flange be made except when the area of the holes exceeds 15 percent of the area of the flange. When the area of the holes exceeds 15 percent, only the area in excess of 15 percent is deducted.

Web holes.   The AISC *Specifications* provide that fastener holes in the web may be neglected. On the other hand, large holes in the web generally require reinforcement and possible stiffeners around the holes.

Example 5-10   *Effect of Holes in Beams*

A W14 × 26 beam has two holes for $\frac{7}{8}$-in.-diameter bolts in each flange. Determine the reduced section modulus.

Solution.   We follow the common practice here of subtracting the same area from both the tension and compression flange whether or not holes are present. Thus the location of the neutral axis is not shifted by holes. Properties of the W14 × 26 are shown in Fig. 5-9. The flange is 5.025 in. wide. Fifteen percent of 5.025 in. is 0.754 in. The bolt holes remove $2(\frac{7}{8} + \frac{1}{8}) = 2.0$ in. from the flange

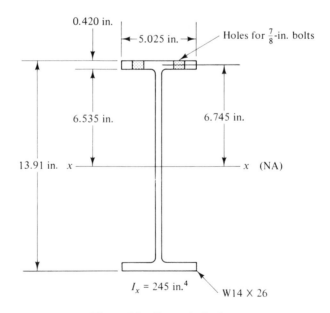

Figure 5-9   Example 5-10.

width. The excess over 15 percent removed is $(2.0 - 0.754) = 1.246$ in. Therefore, the hole area to be removed is $1.246(0.420) = 0.523$ in.^{2} for both the tension and compression flange. The reduced moment of inertia

$$I = 245 - 2(0.523)(6.745)^2 = 197.4 \text{ in.}^4$$

and the section modulus

$$S = \frac{I}{c} = \frac{197.4}{6.955} = 28.4 \text{ in.}^3$$

The section modulus was reduced 19.4 percent by the holes.

5-9 WEB BUCKLING

Two types of buckling of the webs of beams can occur: vertical buckling and diagonal buckling.

Vertical buckling can occur only at points where concentrated loads or reactions are applied. The loads or reactions cause critical compressive stress in a vertical direction near the center of the beam web and the web buckles like a column, as shown in Fig. 5-10(a). Vertical buckling depends on the

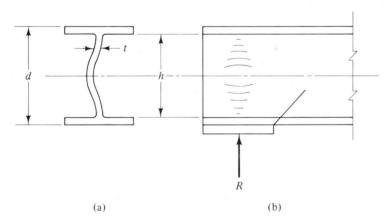

(a)   (b)

Figure 5-10   Vertical buckling of a beam web.

slenderness ratio for a vertical strip of web. The slenderness ratio can be expressed in terms of the depth-to-thickness ratio of the web, d/t_w. However, the proportions of the usual rolled steel sections are such that vertical web buckling does not occur. In deep built-up sections with thin webs vertical buckling is possible (see Sec. 5-6 and AISC Specification 1.5.1.4.1).

Diagonal buckling is the result of shear stresses in the web. The element of a beam web illustrated in Fig. 5-11(a) is assumed to be in a region where

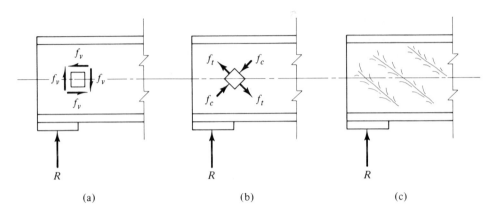

Figure 5-11 (a) Shear stresses acting on an element of beam web; (b) equivalent compression and tensile stresses acting on an element of beam web; (c) diagonal web buckling.

the bending stresses are zero. Recall from strength of materials that as a direct consequence of equilibrium the shear stresses on all four faces of the element must be equal and directed as shown. Recall further that whenever pure shearing stresses act on an element, tensile stresses exist along one diagonal and compressive stresses along the other diagonal of the element as shown in Fig. 5-11(b). If the compressive stresses along the diagonal are large enough, the web will buckle forming a series of wrinkles perpendicular to the direction of the compressive stress [Fig. 5-11(c)]. As in the case of vertical buckling, diagonal buckling does not occur for the usual rolled steel section.

For built-up sections it is necessary to consider diagonal buckling. Diagonal buckling can be controlled by reducing the depth-to-thickness ratio of the web, keeping the shearing stresses low, or by using web stiffeners. Such topics are considered in the design of plate girders.

5-10 WEB CRIPPLING

A load applied over a short length of beam can cause failure due to high compressive stress in the web of the beam above or below the load. The AISC *Specifications* assume that a bearing plate of length N distributes the load over a length of web $(N + k)$ [Fig. 5-12(a)]. Thus the web area in compression at a load is

$$A_c = (N + k)t_w$$

and the allowable compressive stress on the web of a rolled shape at the toe of the fillet from AISC Specification 1.5.1.3.5 is

$$F_a = 0.75F_y$$

At the reaction the compressive stress in the web $f_a = R/A_c$. Therefore, for

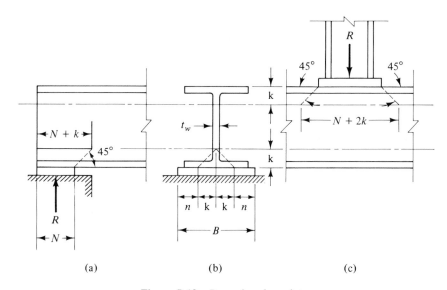

Figure 5-12   Beam bearing plate.

an end reaction we have AISC Formula (1.10-8),

$$\frac{R}{(N+k)t_w} \le 0.75 F_y \tag{5-9}$$

The load at an interior support is assumed to be distributed over a length of web $(N + 2k)$ [Fig. 5-12(c)]. Therefore, at an interior load we have AISC Formula (1.10-9),

$$\frac{R}{(N+2k)t_w} \le 0.75 F_y \tag{5-10}$$

where R = concentrated load or reaction, k
        t_w = thickness of the web, in.
        N = length of bearing plate, in.
        k = distance from toe of fillet to outside of flange, in.

Example 5-11   *Bearing Plate Length*

Determine the length of bearing plate required to prevent failure due to crippling of the web of a W10 × 88 beam if the reaction R = 90 k. Use A36 steel.

Solution.   Properties of the W10 × 88: t_w = 0.605 in., k = $1\frac{5}{8}$ = 1.625 in. Solving for N in Eq. (5-9), we have

$$N \ge \frac{R}{t_w 0.75 F_y} - k = \frac{90}{0.605(0.75)(36)} - 1.625$$

$$\ge 3.88 \text{ in.}$$

Use N = 4.0 in. or larger.

The bearing length is also affected by the strength of the masonry under the plate. This will be discussed in the design of beam bearing plates (Sec. 5-12).

5-11 BEAM DEFLECTION LIMITATIONS

In AISC Specification 1.13.1 limitations are placed on the live load deflection of beams and girders that support plastered ceilings to not more than $\frac{1}{360}$ of the span. This limitation, based on past experience, is intended to prevent cracks in plastered surfaces.

An alternative method of limiting the deflection is given in the Commentary on AISC Specification 1.13.1, where it is suggested that the depth-to-span ratio for fully stressed beams should be not less than $F_y/800$, that is,

$$\frac{d}{L} \geq \frac{F_y}{800}$$

If a smaller ratio is used, the bending stress should be decreased in the same ratio as the depth-to-span ratio is decreased from the above. For fully stressed purlins the suggested depth-to-span ratio should be not less than $F_y/1000$, that is,

$$\frac{d}{L} \geq \frac{F_y}{1000}$$

except in the case of flat roofs. For flat roofs special provisions must be made for ponding—the retention of water due to the deflection of flat roof framing.

For a uniformly loaded beam with a bending stress $f_b = F_b = 0.6F_y$, the suggested depth-to-span ratio of $F_y/800$ corresponds with a deflection $\frac{1}{290}$ of the span and the suggested ratio $F_y/1000$ corresponds with a deflection $\frac{1}{232}$ of the span.

Another method of limiting the deflection is given in the Commentary on AISC Specification 1.13.2, where the depth of a steel beam supporting large open areas free of sources of damping should be no less than $\frac{1}{20}$ of the span to minimize transient vibrations.

Example 5-12 *Deflection Limitations on Beams*

A simply supported beam has a span and loads as shown. Select a W section made of A36 steel that can safely support the loads. Assume full lateral supports. Check the deflection to see if it exceeds $\frac{1}{360}$ of the span and satisfies the limitations given by the AISC *Specifications*, Commentary 1.13.1 and 1.13.2.

Solution. Neglect the weight of the beam. The shear and moment diagrams are shown in Fig. 5-13. For a compact section $F_b = 24$ ksi. The required section modulus $S = M/F_b = 316.8(12)/24 = 158.4$ in.^{3}. From the design selection table try a W24 × 76: $S = 176$ in.^{3}, $d = 23.92$ in., $t_w = 0.440$ in., $I = 2100$ in.^{4}. The section is compact, $F'_y > 36$ ksi.

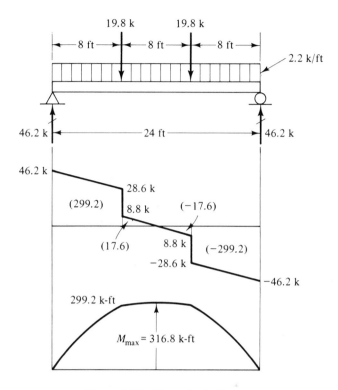

Figure 5-13   Example 5-12.

Check of shear stress

$$f_v = \frac{V}{dt_w} = \frac{46.2}{23.92(0.440)} = 4.39 \text{ ksi} < F_v = 14.4 \text{ ksi} \qquad \text{OK}$$

Check of deflection (deflection = $\frac{1}{360}$ span). From Beam Diagrams and Formulas, cases 1 and 9, AISC *Manual*, for the distributed load $w = 2.2$ k/ft and $E = 29,000$ ksi,

$$d_{max\,1} = \frac{5wL^4}{384EI} = \frac{5(2.2)(24)^4(1728)}{384(29,000)(2100)} = 0.270 \text{ in.}$$

and for the two concentrated loads $P = 19.8$ k,

$$d_{max\,2} = \frac{Pa(3L^2 - 4a^2)}{24EI} = \frac{19.8(8)[3(24)^2 - 4(8)^2](1728)}{24(29,000)(2100)}$$

$$= 0.276 \text{ in.}$$

Note: The conversion factor 1728 *converts* ft^3 *into* in.^{3}. Using superposition, the deflection at the middle of the beam is determined by adding the deflections due to each set of loads.

$$d_{max} = d_{max\,1} + d_{max\,2} = 0.270 + 0.276 = 0.546 \text{ in.}$$

Comparing the maximum deflection with a deflection equal to $\frac{1}{360}$ span, we have

$$\frac{L}{360} = \frac{24(12)}{360} = 0.8 \text{ in.} > d_{max} = 0.546 \text{ in.} \qquad \text{OK}$$

Check of deflection (AISC Specification Commentary 1.13.1)

$$\frac{d}{L} = \frac{23.92}{24(12)} = 0.0831 > \frac{F_y}{800} = \frac{36}{800} = 0.045 \qquad \text{OK}$$

Check of deflection (AISC Specification Commentary 1.13.2)

$$d = 23.92 \text{ in.} > \frac{L}{20} = \frac{23(12)}{20} = 14.4 \text{ in.} \qquad \text{OK}$$

Example 5-13   *Design of Beams and Girders for Floor System*

Beams and girders are used to support a reinforced 4-in. concrete floor as shown in Fig. 5-14. Beams, girders, and floor continue on all sides and the floor provides full lateral support. Assume that all members are simply supported. Reinforced concrete weighs 150 lb/ft³. Use A36 steel to design beams B_1 and girder G_1 if the floor supports a live load of 70 psf and the following limitations on deflection apply.

(a) No limitations are placed on the deflection.

(b) Deflection limitations from the AISC Specification Commentary 1.13.1 apply.

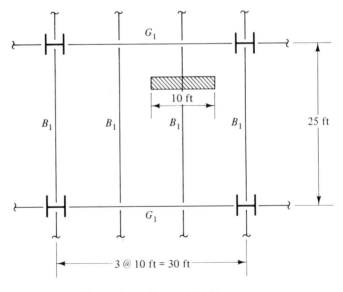

Figure 5-14   Example 5-13.

Solution.   (a) No limitations on deflection.

Beam B_1. Each beam supports a strip of floor 10 ft wide as shown in the figure.

Dead load per linear foot of beam:

$$\text{Concrete floor: } \frac{4}{12}(150)\frac{\text{lb}}{\text{ft}^2}(10 \text{ ft}) = 500 \text{ lb/ft}$$

Live load per linear foot of beam:

$$70\frac{\text{lb}}{\text{ft}^2}(10 \text{ ft}) = \underline{700 \text{ lb/ft}}$$

Total load on beam = 1200 lb/ft

The beam is simply supported and acted on by a uniform load of $w = 1200 \text{ lb/ft} = 1.2 \text{ k/ft}$. The beam, shear diagram, and moment diagram are shown in Fig. 5-15.

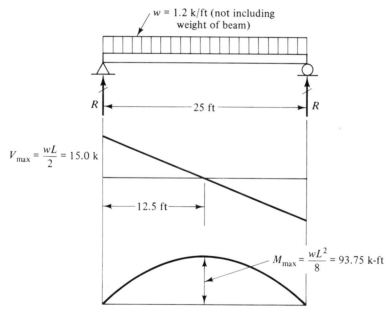

Figure 5-15   Example 5-13.

Assume a compact section: $F_b = 24$ ksi. The required section modulus $S = M/F_b = 93.75(12)/24 = 46.9 \text{ in.}^3$. From the design selection table, try a W14 × 34: $S_x = 48.6 \text{ in.}^3$. The section is compact; $F_y' > 36$ ksi. The maximum moment including the weight of the beam

$$M_{\text{max}} = 93.75 + \frac{0.034(25)^2}{8} = 96.41$$

and the required section modulus

$$S = \frac{M}{F_b} = \frac{96.41(12)}{24} = 48.2 \text{ in.}^3 < S_x = 48.6 \text{ in.}^3 \qquad \text{OK}$$

Check of shear stress

$$V_{max} = \frac{wL}{2} = \frac{1.234(25)}{2} = 15.42 \text{ k}$$

$$f_v = \frac{V}{dt_w} = \frac{15.42}{13.98(0.288)} = 3.87 \text{ ksi} < 0.4F_y = 14.4 \text{ ksi} \qquad \text{OK}$$

$$R_L = R_R = \frac{wL}{2} = 15.42 \text{ k}$$

Use a W14 × 34 for beam B_1.

Girder G_1. Each girder supports the end reactions from four beams—two end reactions at each third point of the girder—and its own weight. The girder, shear diagram, and moment diagram are shown in Fig. 5-16. Assume a compact

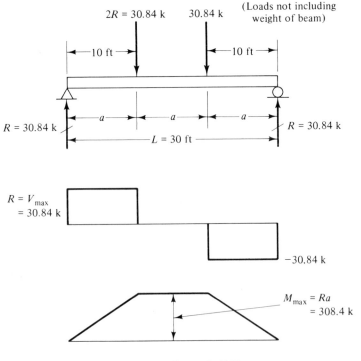

Figure 5-16   Example 5-13.

section; $F_b = 24$ ksi. The required section modulus

$$S = \frac{M}{F_b} = \frac{308.4(12)}{24} = 154.2 \text{ in.}^3$$

From the allowable design selection table, try a W24 × 76: $S_x = 176 \text{ in.}^3$ and the section is compact. The maximum moment including the weight of the beam

$$M_{\max} = 308.4 + \frac{0.076(30)^2}{8} = 317.0 \text{ in.}^3$$

and the required section modulus

$$S = \frac{M}{F_b} = \frac{317(12)}{24} = 158.5 \text{ in.}^3 < S_x = 176 \text{ in.}^3 \qquad \text{OK}$$

Check of shear stress

$$V_{\max} = 30.84 + \frac{0.076(30)}{2} = 32.0 \text{ k}$$

$$f_v = \frac{V}{dt_w} = \frac{32.0}{23.92(0.440)} = 3.04 \text{ ksi} < 0.4F_y = 14.4 \text{ ksi} \qquad \text{OK}$$

Use a W24 × 76 for girder G_1.

(b) Deflection limitations from AISC Specification Commentary 1.13.1.

Beam B_1

$$d = 13.98 \text{ in.} \geq \frac{F_y}{800} L = \frac{36}{800}(25)(12) = 13.5 \text{ in.} \qquad \text{OK}$$

Use W14 × 34 for beam B_1.

Girder G_1

$$d = 23.96 \text{ in.} \geq \frac{F_y}{800} L = \frac{36}{800}(30)(12) = 16.2 \text{ in.} \qquad \text{OK}$$

Use W24 × 76 for girder G_1.

5-12 DESIGN OF BEAM BEARING PLATES

Steel beam bearing plates are used under beams to distribute the beam reaction over sufficient length of the beam so that the web is not overstressed and over sufficient area of the supporting footing so that the footing is not overstressed.

The design procedure given here follows that of the AISC *Manual*. The method assumes that the beam reaction is distributed to the beam bearing plate over a rectangle N by $2k$. The bearing plate then distributes this load uniformly to the footing under the plate. The plate has a tendency to bend in a single plane perpendicular to the beam [Fig. 5-12(b)]. The thickness of the base plate is determined by the resulting bending moment. The thickness of the plate may be determined by Eq. (4-13), the same one used to find the thickness of a column base plate.

$$t = \sqrt{\frac{3f_c n^2}{0.75 F_y}}$$

where $n = (B - 2k)/2$, in.

f_c = actual bearing stress, ksi

Example 5-14

Design a bearing plate of A36 steel for a W21 × 93 beam with an end reaction of 70 k. The length of the bearing plate N is limited to 10 in. and the masonry has an allowable stress $F_p = 0.3$ ksi.

Solution.   Properties for the W21 × 93: $t_w = 0.580$ in., $k = 1\frac{11}{16} = 1.688$ in. To find the minimum bearing length N, based on crippling, we solve for N in Eq. (5-9):

$$N = \frac{R}{0.7 F_y t_w} - k = \frac{70}{0.75(36)(0.580)} - 1.688$$

$$= 2.78 \text{ in.} < 10 \text{ in.} \qquad \text{OK}$$

The required plate area

$$A = \frac{P}{F_p} = \frac{70}{0.3} = 233.3 \text{ in.}^2$$

Therefore, the width

$$B = \frac{A}{N} = \frac{233.3}{10} = 23.3 \text{ in.} \quad (\text{use } 24 \text{ in.})$$

The actual bearing stress

$$f_b = \frac{R}{BN} = \frac{70}{24(10)} = 0.292 \text{ ksi}$$

Then $n = (B - 2k)/2 = [24 - 2(1.688)]/2 = 10.31$ in. and the plate thickness

$$t = \sqrt{\frac{3f_c n^2}{0.75 F_y}} = \sqrt{\frac{3(0.292)(10.31)^2}{0.75(36)}} = 1.86 \text{ in.} \quad (\text{use } 1\frac{7}{8} \text{ in.})$$

Use bearing plate PL$1\frac{7}{8}$ × 10 × 2 ft-0 in.

5-13 LATERAL BUCKLING OF BEAMS

The failure of columns is usually due to instability. For beams that are not continuously braced, instead of failure due to bending, the compression flange of the beam may become unstable. The tension stresses in the opposite flange tend to keep the beam straight. Consequently, beam buckling is a combination of lateral bending and torsion or twisting of the cross section.

The AISC specifications are based on the solution of the lateral-torsional buckling of a doubly symmetric beam acted on by bending couples at each end that produce *pure bending* in the plane of the web. The bending is assumed to be in the *elastic range* and that buckling occurs without distortion of the cross section.

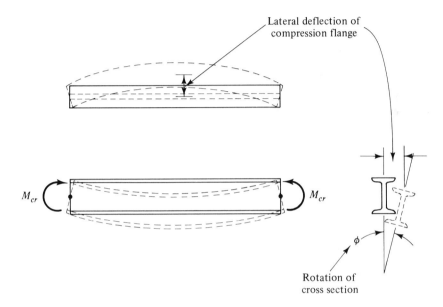

Figure 5-17   Lateral buckling of a simply supported beam acted on by a uniform bending moment.

In Fig. 5-17 we consider such a beam. As the bending couples increase the beam deflects vertically. When the bending couples reach the critical value M_{cr} the cross section twists, the compression flange moves laterally, and the cross section rotates through an angle. The critical moment M_{cr} is made up of two parts: the torsional resistance and the warping resistance. They can be combined in the following form:

$$M_{cr} = \sqrt{\left(\frac{\text{torsional}}{\text{resistance}}\right)^2 + \left(\frac{\text{warping}}{\text{resistance}}\right)^2} \qquad (5\text{-}11)$$

The *torsional resistance*, which is commonly called St.-Venant torsion, occurs when equal and opposite torsional couples act at the unrestrained ends of a member. All of the cross sections of the member warp of tilt freely by an equal amount. When a wide flange beam buckles laterally the warping of the flanges is partially restrained. Bending moments and shears are developed in the flanges and give rise to the *warping resistance* of the beam.

If the critical moment given by Eq. (5-11) is divided by the section modulus S_x, the critical stress F_{cr} is obtained. Dividing the critical stress by a basic factor of safety of $\frac{5}{3}$, an equation for allowable stress F_b is obtained.

$$F_b = \sqrt{\left[\frac{12(10)^3}{Ld/A_f}\right]^2 + \left[\frac{170(10)^3}{(L/r_T)^2}\right]^2}$$ (5-12)

where L = distance between lateral supports, in.

d = depth of the section, in.

A_f = area of the compression flange, in.^{2}

r_T = radius of gyration of the compression flange plus one-third of the compression web taken about the y axis of the cross section, in.

Equation (5-12) is applicable to a simply supported beam acted on by end couples (uniform bending moment) when the buckling is elastic.

Multiplying the equation by a bending coefficient C_b to compensate for the difference between the actual bending moments and the uniform bending moments assumed in the development of the equation, we have

$$F_b = \sqrt{\left[\frac{12(10)^3 C_b}{Ld/A_f}\right]^2 + \left[\frac{170(10)^3 C_b}{(L/r_T)^2}\right]^2}$$ (5-13)

The bending coefficient C_b ranges in value from 1.0 to 2.3. Specific values of the coefficients are discussed with Eq. (5.18).

In Eq. (5-13) the first term represents the torsional resistance and the second term the warping resistance. For shallow thick-walled beams the torsional resistance predominates and the warping resistance can be neglected and the allowable bending stress

$$F_b = \frac{12(10)^3 C_b}{Ld/A_f}$$ (5-14)

For deep thin-walled beams the warping resistance predominates and the torsional resistance can be neglected. Thus the allowable bending stress

$$F_b = \frac{170(10)^3 C_b}{(L/r_T)^2}$$ (5-15)

AISC Allowable Bending Stress

AISC Specification 1.5.1.4.5(2) follows a conservative approach by using the larger value of stress obtained from the torsional resistance and warping resistance [Eqs. (5-14) and (5-15)] for the allowable stress. Both equations are for elastic buckling only. However, for short beams the warping resistance results in stresses in the plastic or inelastic range. Therefore, rather than a Euler-type equation like Eq. (5-15) a parabola is used for allowable bending stresses greater than one-third of the yield stress. For torsional resistance, Eq. (5-14) is used for both elastic and plastic stresses. Tests indicate that any error in the plastic range will be small.

AISC Specification 1.5.1.4.5(2) states that the compression bending stress F_b for members having an axis of symmetry and loading in the plane of their web and compression on extreme fibers of channels bent about their major (x axis) must be the larger of the following equation but not greater than $0.6F_y$. When

$$\sqrt{\frac{102(10)^3 C_b}{F_y}} \le \frac{L}{r_T} \le \sqrt{\frac{510(10)^3 C_b}{F_y}}$$

$$F_b = \left[\frac{2}{3} - \frac{F_y(L/r_T)^2}{1530(10)^3 C_b} \right] F_y \tag{5-16}$$

$$\text{[AISC Formula (1.5-6a)]}$$

or when

$$\frac{L}{r_T} \ge \sqrt{\frac{510(10)^3 C_b}{F_y}}$$

$$F_b = \frac{170(10)^3 C_b}{(L/r_T)^2} \tag{5-17}$$

$$\text{[AISC Formula (1.5-6b)]}$$

When the compression flange is solid and approximately rectangular in cross section and its area is not less than that of the tension flange,

$$F_b = \frac{12(10)^3 C_b}{Ld/A_f} \tag{5-18}$$

$$\text{[AISC Formula (1.5-7)]}$$

where L = unbraced length of the compression flange, in.

r_T = radius of gyration of the compression flange plus one-third of the compression web taken about the y axis of the cross section, in.

A_f = area of the compression flange, in.^{2}

$C_b = 1.75 + 1.05(M_1/M_2) + 0.3(M_1/M_2)^2 \le 2.3$, where M_1 is the numerically smaller and M_2 the numerically larger bending moment at the braced ends of an unbraced length. If the end moments produce single curvature, the ratio of M_1 to M_2 is negative. If they produce reverse curvature, the ratio is taken as positive. When the bending moment at any point within an unbraced length is larger than that at both ends, C_b is taken as unity.

Graphs for Allowable Bending Stress

For $C_b = 1.0$ and $F_y = 36$ ksi, graphs of the allowable stress F_b versus L/r_T from AISC Formulas (1.5-6a) and (1.5-6b) are shown in Fig. 5-18(a) and the allowable bending stress F_b versus Ld/A_f from AISC Formula (1.5-7) is shown in Fig. 5-18(b).

The maximum unbraced length for which $F_b = 0.6F_y$ can be obtained by setting Eqs. (5-16) and (5-18) equal to $0.6F_y$ and solving for L. With $C_b = 1$ in the resulting equations the maximum unbraced length called L_u is the larger

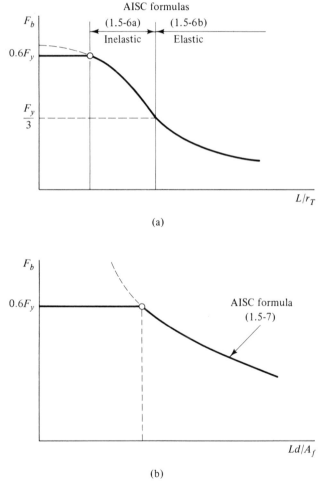

(a)

(b)

Figure 5-18 Allowable bending stress equations for beams without continuous lateral supports: (a) warping resistance, A36 steel, $C_b = 1.0$; (b) torsional resistance, A36 steel, $C_b = 1.0$.

of the following:

$$L = \frac{20(10)^3}{12(d/A_f)F_y} \text{ ft} \quad \text{or} \quad L = \frac{r_T}{12}\sqrt{\frac{102(10)^3}{F_y}} \text{ ft}$$

Values of L_u and also of L_c (Sec. 5-6) are given in the allowable stress design selection table for beams of the AISC *Manual.*

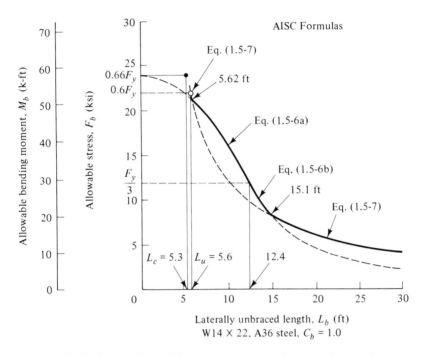

Figure 5-19 Comparison of allowable stress equations for a beam without continuous lateral supports.

TABLE 5-1   Summary of Fig. 5-19

Unbraced Length range, L_b (ft)	Allowable Bending Stress or AISC Formula	Comment
0.0–5.3	$0.66F_y$	$L_b = L_c = 5.3$ ft
5.3–5.6	$0.60F_y$	$L_c = 5.3$ ft $\leq L_b \leq L_u = 5.6$ ft
5.6–5.62	1.5-7	Torsion
5.62–12.4	1.5-6(a)	Warping (inelastic)
12.4–15.1	1.5-6(b)	Warping (elastic)
15.1–	1.5-7	Torsion

A comparison of AISC Formulas (1.5-6a), (1.5-6b), and (1.5-7) for a W14 × 22 beam with $C_b = 1.0$ and $F_y = 36$ ksi appears in Fig. 5-19. A summary of the figure is given in Table 5-1.

Example 5-15 *Allowable Bending Stresses and Moments for Beams Without Lateral Supports*

Lateral supports are provided at the ends of a simply supported W33 × 130 beam of A36 steel. Determine the allowable bending stress and allowable bending moment for the beam if it has spans L_b as follows: **(a)** 10 ft, **(b)** 13 ft, **(c)** 14 ft, **(d)** 20 ft, and **(e)** 30 ft.

Solution. Properties of a W33 × 130: $L_c = 12.1$ ft, $L_u = 13.8$ ft, $S_x = 406$ in.^{3}, $r_T = 2.88$ in., $d/A_f = 3.36$ in.$^{-1}$.

    (a) $L_b = 10$ ft,

$$L_b = 10 \text{ ft} < L_c = 12.1 \text{ ft}$$

Therefore,

$$F_b = 0.66F_y = 0.66(36) = 24 \text{ ksi}$$

and

$$M_b = \frac{F_b S_x}{12} = \frac{24(406)}{12} = 812 \text{ k-ft}$$

    (b) $L_b = 13$ ft,

$$L_c = 12.1 \text{ ft} < L_b = 13 \text{ ft} < L_u = 13.8 \text{ ft}$$

Therefore,

$$F_b = 0.6F_y = 22 \text{ ksi}$$

Note that a rounded value for F_b from Appendix A, Table 1 of the AISC *Specifications* is used. Thus the allowable bending moment

$$M_b = \frac{F_b S_x}{12} = \frac{22(406)}{12} = 744 \text{ k-ft}$$

    (c) $L_b = 14$ ft,

$$\frac{Ld}{A_f} = 14(12)(3.36) = 564 \qquad \frac{L}{r_T} = \frac{13(12)}{2.88} = 58.3$$

From Eq. (5-18) [AISC Formula (1.5-7)],

$$F_{b1} = \frac{12(10)^3 C_b}{Ld/A_f} = \frac{12(10)^3(1.0)}{564} = 21.3 \text{ ksi}$$

Since

$$\frac{L}{r_T} = 58.3 < \sqrt{\frac{510(10)^3 C_b}{F_y}} = \sqrt{\frac{510(10)^3(1.0)}{36}} = 119$$

use Eq. (5-16) [AISC Formula (1.5-6a)].

$$F_{b2} = \left[\frac{2}{3} - \frac{F_y(L/r_T)^2}{1530(10)^3 C_b}\right] F_y$$

For $F_y = 36$ ksi,

$$F_{b2} = 24 - \frac{(L/r_T)^2}{1181 C_b} = 24 - \frac{(58.3)^2}{1181(1.0)} = 21.1 \text{ ksi}$$

The control value is $F_{b1} = 21.3$ ksi; therefore,

$$M_b = \frac{F_b S_x}{12} = \frac{21.3(406)}{12} = 721 \text{ k-ft}$$

(d) $L_b = 20$ ft,

$$\frac{Ld}{A_f} = 20(12)(3.36) = 806 \qquad \frac{L}{r_T} = \frac{20(12)}{2.88} = 83.3$$

From Eq. (5.18) [AISC Formula (1.5-7)],

$$F_{b1} = \frac{12(10)^3 C_b}{Ld/A_f} = \frac{12(10)^3(1.0)}{806} = 14.89 \text{ ksi}$$

Since

$$\frac{L}{r_T} = 83.3 < \sqrt{\frac{510(10)^3 C_b}{F_y}} = 119$$

use Eq. (5-16) [AISC Formula (1.5-6a)].

$$F_{b2} = 24 - \frac{(L/r_T)^2}{1181 C_b} = 24 - \frac{(83.3)^2}{1181(1.0)} = 18.12 \text{ ksi}$$

The control bending stress is $F_{b2} = 18.12$ ksi; therefore,

$$M_b = \frac{F_b S_x}{12} = \frac{18.12(406)}{12} = 613 \text{ k-ft}$$

(e) $L_b = 30$ ft,

$$\frac{Ld}{A_f} = 30(12)(3.36) = 1210 \qquad \frac{L}{r_T} = \frac{30(12)}{2.88} = 125$$

From Eq. (5-18) [AISC Formula (1.5-7)],

$$F_{b1} = \frac{12(10)^3 C_b}{Ld/A_f} = \frac{12(10)^3(1.0)}{1210} = 9.92 \text{ ksi}$$

Since

$$\frac{L}{r_T} = 125 > \sqrt{\frac{510(10)^2 C_b}{F_y}} = 119$$

use Eq. (5-17) [AISC Formula (1.5-6b)].

$$F_{b2} = \frac{170(10)^3 C_b}{(L/r_T)^2} = \frac{170(10)^3(1.0)}{(125)^2} = 10.88 \text{ ksi}$$

The control stress $F_{b2} = 10.88$ ksi; therefore,

$$M_b = \frac{F_b S_x}{12} = \frac{10.88(406)}{12} = 368 \text{ k-ft}$$

5-14 DESIGN OF BEAMS WITHOUT CONTINUOUS LATERAL BRACING

The allowable moments in beams charts of the AISC *Manual* will be used to design beams that are not continuously braced. The curves for the charts were constructed in the same way as Fig. 5-19. The allowable stress F_b was multiplied by the section modulus of the beam to obtain the allowable bending moment. The total allowable moment in k-ft was plotted with respect to the unbraced length in feet.

The value of unbraced length corresponding to L_c is shown on the beam curves of the charts as a solid circle or dot, and the unbraced length corresponding to L_u is shown as an open circle. The bending coefficient C_b was taken as 1.0.

Use of the charts is illustrated in the following examples.

Example 5-16   *Simply Supported Beam, Uniform Load, $C_b = 1.0$*

Select the lightest W section made of A36 steel for a simply supported beam with a uniform load of 3.0 k/ft over the entire span of 25 ft if lateral supports are provided at the ends of the beam only.

Solution.   The bending moments at the ends of the beam are both zero and the maximum bending moment at the middle of the beam $M_{max} = wL^2/8 = 3.0(25)^2/8 = 234$ k-ft. From AISC Specification 1.5.1.4.5(a): "When the bending moment at any point within an unbraced length is larger than at both ends of this length, the value of C_b shall be taken as unity."

Entering the allowable moments in beams chart in the AISC *Manual*, p. 2-58, at an unbraced length of 25 ft we move up in the chart until we reach an allowable bending moment of 234 k-ft. Any beam curve that appears directly above that point has an allowable bending moment of at least 234 k-ft for an unbraced length of 25 ft. The first solid beam curve above that point is for the most economical section by weight. Moving up, we find the following beam curves in sequence: W24 × 94, W27 × 94, and the first solid beam curve for the W18 × 86. The W18 × 86 has an allowable moment $M_b = 257$ k-ft. The additional moment capacity of that beam is more than adequate to support the additional moment due to the beam weight. Use a W18 × 86.

Example 5-17 *Simply Supported Beam, Concentrated Load, and Lateral Supports at the Ends and Middle of the Span*

Select the lightest W section made of A36 steel for a simply supported beam with a concentrated load of 36.5 k at the middle of a 28-ft span if lateral supports are provided at the ends and middle of the span. For the bending coefficient:

(a) Assume that $C_b = 1.0$.
(b) Calculate values of C_b by formula.

Solution. (a) We calculate the reactions and draw the shear and moment diagram in Fig. 5-20. With $C_b = 1.0$ enter the beam chart on p. 2-58 of the

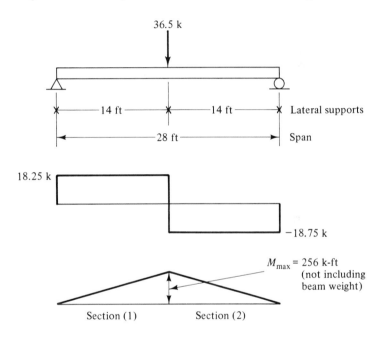

Figure 5-20   Example 5-17.

AISC *Manual* with an unbraced length of 14 ft and a bending moment of 256 k-ft. As we move up the chart, the first solid line is for a W18 × 76. The W18 × 76 has an allowable moment of $M_b = 267.5$ k-ft. The reserve moment capacity will be adequate to support the additional moment due to the beam weight. Use a W18 × 76.

(b) The bending moments on the ends of section (1) of the beam (Fig. 5-20) are $M_1 = 0$ and $M_2 = 256$ k-ft; therefore, $M_1/M_2 = 0$. From the bending coefficient equation, $C_b = 1.75$.

An inspection of the AISC formulas shows that the allowable moments in beams chart can be used for a $C_b = 1.75$ if L_b is replaced by an effective length L_b/C_b when AISC Formula (1.5-7) applies and if L_b is replaced by an effective length $L_b/\sqrt{C_b}$ when AISC Formula (1.5-6a) or (1.5-6b) applies. This method does not apply when $F_b = 0.66F_y$ (when the effective length is less than L_c). However, it does apply when $F_b = 0.6F_y$.

Since most of the curves on the chart are based on AISC Formula (1.5-7), we enter the chart on p. 2-58 of the AISC *Manual* with an effective length of $L_b/C_b = 14/1.75 = 8$ ft and a maximum moment of $M_b = 256$ k-ft. The first solid line we come to is for a W21 × 68. (The curve does not apply—$L_b/C_b = 8$ ft $< L_c = 8.7$ ft.) As we continue upward, moving from p. 2-58 to p. 2-57, the second solid line is for a W24 × 68. The curve does not apply in that region. However, if we follow the curve downward, we see that the curve does apply in the region where $L_b = 14$ ft $> L_c = 9.5$ ft and the beam has an allowable bending moment of $M_b = 282$ k-ft.

Check of W24 × 68. Properties of W24 × 68: $r_T = 2.26$ in., $d/A_f = 4.52$ in.$^{-1}$, $S_x = 154$ in.^{3}. For $L_b = 14$ ft, $C_b = 1.75$,

$$\frac{Ld}{A_f} = 14(12)(4.52) = 759 \qquad \frac{L}{r_T} = \frac{14(12)}{2.26} = 74.3$$

From Eq. (5-18) [AISC Formula (1.5-7)],

$$F_{b1} = \frac{12(10)^3 C_b}{Ld/A_f} = \frac{12(10)^3(1.75)}{759}$$

$$= 27.7 \text{ ksi} > 0.6F_y = 22 \text{ ksi}$$

Use $F_b = 22$ ksi; therefore,

$$M_b = \frac{F_b S_x}{12} = \frac{22(154)}{12} = 282 \text{ k-ft}$$

The maximum bending moment including the weight of the beam

$$M_{\max} = 256 + \frac{0.068(28)^2}{8} = 263 \text{ k-ft} < M_b = 282 \text{ k-ft} \qquad \text{OK}$$

Use a W24 × 68.

Example 5-18   *Cantilever Beam, $F_y = 50$ ksi*

Select the most economical section for the cantilever beam with lateral supports as shown in Fig. 5-21. Use A572 Grade 50 steel.

Solution. The shear and moment diagrams are illustrated in Fig. 5-21. With $C_b = 1.0$ we enter the allowable moments in beams chart ($F_y = 50$ ksi), p. 2-69 of the AISC *Manual*, with $L_b = 20$ ft and $M_{\max} = 400$ k-ft. A check of the W18 × 97 shows that it is not adequate. A check of the next section, a W30 × 99, shows that it is adequate. Use a W30 × 99:

$$M_{\max} = 419.8 \text{ k-ft} < M_b = 426.9 \text{ k-ft} \qquad \text{OK}$$

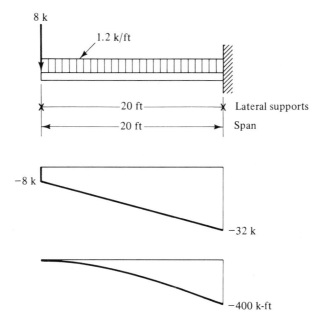

8 k

1.2 k/ft

├──────────────20 ft──────────────┤ Lateral supports

├──────────────20 ft──────────────┤ Span

−8 k

−32 k

−400 k-ft

Figure 5-21   Example 5-18.

Example 5-19   *Beam with Overhang*

Select the most economical section for the beam with loads and lateral supports as shown in Fig. 5-22. Use A36 steel.

Solution.   The moment diagram is shown in Fig. 5-22. Each section of the beam between lateral supports must be considered separately.

Section (1).   The bending moments $M_1 = 0$ and $M_2 = 216$ k-ft. Therefore, $M_1/M_2 = 0$ and the bending coefficient $C_b = 1.75$. We enter the allowable moments in beams chart, p. 2-54 of the AISC *Manual*, with an effective length $L_b/C_b = 20/1.75 = 11.4$ ft. A check of the W21 × 62 shows that it is not adequate. A check of the next section, a W21 × 68, shows that it is adequate.

Section (2).   The bending coefficient

$$C_b = 1.75 + 1.05 \frac{168}{216} + 0.3 \left(\frac{168}{216}\right)^2 = 2.75 \leq 2.3$$

Therefore, use $C_b = 2.3$.

Entering the table as before with an effective length $L_b/C_b = 20/2.3 = 8.7$ ft, we see that the W21 × 68 is adequate since it can support a bending moment of approximately 256 k-ft.

Section (3).   The bending coefficient is again $C_b = 1.75$ [see section (1)]. Entering the table with an effective length $L_b/C_b = 14/1.75 = 8$ ft, we see that

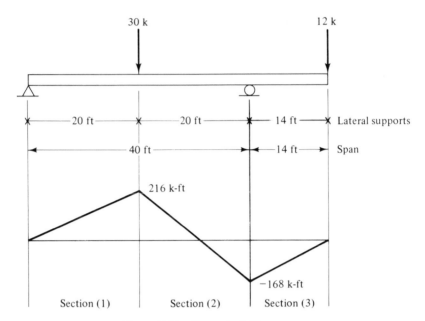

Figure 5-22   Example 5-19.

the W21 × 68 is adequate since it can support a bending moment of approximately 256 k-ft.

Check. Although not necessary, a check was made of the W21 × 68 beam for each section. The maximum moments include the beam weight of 0.068 k/ft. For sections (1), (2), and (3) the allowable bending stress F_p = 22 ksi and the allowable bending moment M_b = 256.7 k-ft. For sections (1) and (2) the maximum bending moment M_{max} = 226.3 k-ft and for section (3) the maximum bending moment M_{max} = 174.7 k-ft. Use a W21 × 68.

Example 5-20   *Continuous Beam with Dead Loads*

A continuous beam has two equal 18-ft spans and a uniform dead load of 4.05 k/ft over each span. Select the most economical beam made of A36 steel if lateral supports are provided at intervals of **(a)** 6 ft and **(b)** 18 ft.

Solution.   Assume a beam weight of 50 lb/ft. Therefore, w = 4.1 k/ft. The beam with load, spans, and lateral supports is shown in Fig. 5-23. The beam reactions are determined by superposition from the formulas given on p. 2-124 of the AISC *Manual,* number 29:

$$R_L = R_1 + R_3 = \frac{7}{16}wL - \frac{1}{16}wL = \frac{3}{8}wL$$

$$= \frac{3(4.1)(18)}{8} = 27.68 \text{ k}$$

$$R_M = 2R_2 = 2\left(\frac{5}{8}\right)wL = \frac{5(4.1)(18)}{4} = 92.25 \text{ k}$$

$$R_R = R_L = 27.68 \text{ k}$$

The shear and moment diagrams illustrated in Fig. 5-23 were drawn from the loads and reactions.

(a) $L_b = 6$ ft. Assume a compact section and adequate bracing. The moments may be adjusted as indicated in the last paragraph of AISC Specification 1.5.1.4.1 as follows.

The maximum negative moment can be decreased by 0.1; therefore,

$$M_{max} = 0.9(-166.0) = -149.4 \text{ k-ft}$$

If the maximum positive moment is increased by 0.1 of the average of the

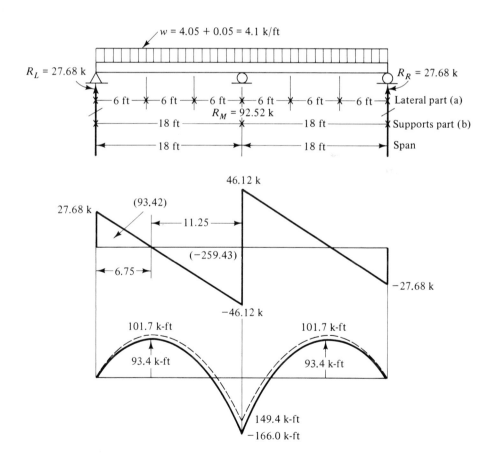

Figure 5-23   Example 5-20.

negative moments,

$$M_{max} = M + \frac{0.1(M_L + M_M)}{2} = 93.4 + \frac{0.1(0 + 166.0)}{2}$$

$$= 101.7 \text{ k-ft}$$

The adjusted moments are shown by the dashed line on the moment diagram. For a compact section with adequate bracing, $F_b = 0.66 F_y = 24$ ksi. The required section modulus $S = M/F_b = 149.4(12)/24 = 74.7 \text{ in.}^3$. From the allowable stress design selection table we select a W21 × 44. The section is compact, $F_y' > 36$ ksi, and adequately braced, $L_b = 6$ ft $< L_c = 6.6$ ft. Use a W21 × 44.

(b) $L_b = 18$ ft. Assume that the bracing is not adequate for moment redistribution. Therefore, we use the original bending moments. The moments for either span $M_1 = 0$ and $M_2 = -166$ k-ft; therefore, $M_1/M_2 = 0$ and $C_b = 1.75$. Entering the allowable moments in beams chart, p. 2-59 of the AISC *Manual*, with an effective length of $L_b/C_b = 18/1.75 = 10.3$ ft, we select a W18 × 55 section that has an allowable bending moment $M_b = 180$ k-ft. The section is more than adequate. Use a W18 × 55.

5-15 BIAXIAL BENDING OF SYMMETRIC SECTIONS

Biaxial bending is simultaneous bending about both the x and y axes. For biaxial bending of a section with at least one axis of symmetry and loading through the centroid of the cross section, the maximum bending stress is found by superposition. That is,

$$f_b = f_{bx} + f_{by} = \frac{M_x}{S_x} + \frac{M_y}{S_y} \tag{5-19}$$

where the subscripts x and y refer to the axis about which bending occurs.

The bending stress f_b must be less than or at most equal to the allowable value F_b; hence

$$F_b \geq \frac{M_x}{S_x} + \frac{M_y}{S_y} \tag{5-20}$$

Multiplying by S_x and dividing by F_b, we have

$$S_x \geq \frac{M_x + (S_x/S_y)M_y}{F_b} \tag{5-21}$$

To obtain a trial section we assume a value for the ratio S_x/S_y and calculate the required section modulus from Eq. (5-21). We then select several sections from the allowable stress design selection table that have the required section modulus. The sections are then checked to determine the lightest section that is satisfactory. The following examples illustrate the method.

Example 5-21   *Design of Beam with Biaxial Bending*

Select the lightest W or M section to support moments of $M_x = 90$ k-ft and $M_y = 18$ k-ft. Use A36 steel. Assume that adequate bracing has been provided.

Solution.   Values of S_x/S_y range from approximately 3 to 15 with an average of about 6. Assume an average value of 6.0. From AISC Specification 1.5.1.4.5, the maximum allowable stress for biaxial bending $F_b = 0.6F_y = 0.6(36) = 22$ ksi. From Eq. (5-21),

$$S_x \geq \frac{M_x + (S_x/S_y)M_y}{F_b} = \frac{[90 + 6(18)]12}{22} = 108 \text{ in.}^3$$

From the allowable stress design selection table, the following lightest shapes that have a section modulus larger than 108 in.^{3} are selected: W24 × 55, W21 × 62, W24 × 62, and W21 × 68. The first three sections are unsatisfactory.

Check of W21 × 68. $S_x = 140$ in.^{3}, $S_y = 15.7$ in.^{3}. The required section modulus

$$S_x = \frac{M_x(S_x/S_y)M_y}{F_b} = \frac{[90 + (140/15.7)(18)]12}{22}$$

$$= 136.6 \text{ in.}^3 < S_x = 140 \text{ in.}^3 \qquad \text{OK}$$

Use a W21 × 68.

For biaxial bending the interaction equation for a combined axial load and biaxial bending from AISC Formula 1.6-2 can be used. With the stress due to an axial load $f_a = 0$ (no axial load present), the interaction equation reduces to

$$\frac{f_{bx}}{F_{bx}} + \frac{f_{by}}{F_{by}} \leq 1.0 \tag{5-22}$$

The allowable stresses will be taken here from AISC Specifications 1.5.1.4.3 and 1.5.1.4.5, where $F_{bx} = 0.6F_y = 22$ ksi and $F_{by} = 0.75F_y = 27$ ksi. For the W24 × 68: $S_x = 140$ in.^{3}, $S_y = 15.7$ in.^{3}. The bending stresses

$$f_{bx} = \frac{M_x}{S_x} = \frac{90(12)}{140} = 7.71 \text{ ksi}$$

$$f_{by} = \frac{M_y}{S_y} = \frac{18(12)}{15.7} = 13.75 \text{ ksi}$$

Substituting into Eq. (5-22) yields

$$\frac{f_{bx}}{F_{bx}} + \frac{f_{by}}{F_{by}} = \frac{7.71}{22} + \frac{13.75}{27} = 0.350 + 0.509$$

$$= 0.859 < 1.00 \qquad \text{OK}$$

Therefore, the W24 × 68 beam is satisfactory. The W21 × 62 section is also satisfactory by this method.

Example 5-22   *Design of Purlin with Biaxial Bending*

Design a purlin of A36 steel with a slope of 1 to 2 as shown in Fig. 5-24. The purlins are supported by trusses 20 ft apart and sag rods are used at the midpoint between the trusses. Assume that the roof provides lateral support and there is no torsion. The roof weighs 4.93 psf, the snow load is 25 psf on a horizontal projection, and the wind load is 17.7 psf perpendicular to the roof.

(a) Use a purlin depth $\frac{1}{30}$ span.

(b) Use a purlin depth $\frac{1}{24}$ span.

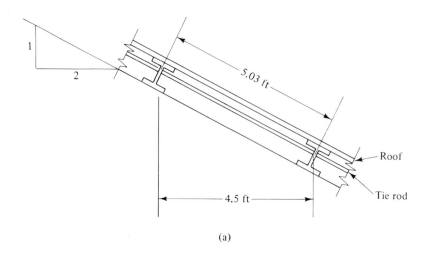

(a)

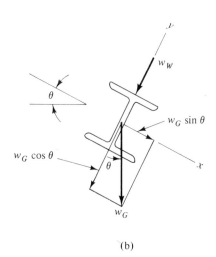

(b)

Figure 5-24   Example 5-22.

Solution

Dead load per linear foot of purlin:

$$\text{Roof: } 4.93 \text{ lb/ft}^2 \,(5.03 \text{ ft}) \quad = \quad 24.8 \text{ lb/ft}$$
$$\text{Purlin (estimate 50\% of roof)} = \underline{\quad 12.4 \text{ lb/ft}}$$
$$\text{Total: } w_D = \quad 37.2 \text{ lb/ft}$$

Snow load per linear foot of purlin:

$$\text{Snow: } 25 \text{ lb/ft}^2 \,(4.5 \text{ ft}) \quad W_S = 112.5 \text{ lb/ft}$$

Wind load per linear foot of purlin:

$$\text{Wind: } 17.7 \text{ lb/ft}^2 \,(5.03 \text{ ft}) \; w_W = \quad 89.0 \text{ lb/ft}$$

Consider two load combinations (see Sec. 2-8):

1. Dead load and full snow load
2. Dead load, wind load, and one-half snow load

For case 1, the gravity load

$$w_G = w_D + w_S = 37.2 + 112.5 = 149.7 \text{ lb/ft}$$

The gravity load will be resolved into components perpendicular to the roof (y component) and parallel to the roof (x component), as illustrated in Fig. 5-24(b).

$$w_y = w_G = 149.7 \cos 26.57° = 133.9 \text{ lb/ft}$$

$$w_x = w_G = 149.7 \sin 26.57° = 67.0 \text{ lb/ft}$$

For case 2, the gravity load w_G and wind load w_W are given by

$$w_G = w_D + \frac{w_S}{2} = 37.23 + \frac{112.5}{2} = 93.5 \text{ lb/ft}$$

$$w_W = 89.2 \text{ lb/ft}$$

The gravity load will be resolved into x and y components and the wind load will be added to the y component.

$$w_y = w_W + w_G \cos \theta = 89.2 + 93.5 \cos 26.57°$$

$$= 172.8 \text{ lb/ft}$$

$$w_x = w_G \sin \theta = 93.5 \sin 26.57° = 41.8 \text{ lb/ft}$$

The allowable stress may be increased by one-third for any load combination that includes wind, or alternatively, the load may be multiplied by three-fourths (Example 4-10); thus

$$w_y = 0.75(172.8) = 129.6 \text{ lb/ft} < 133.9 \text{ lb/ft}$$

$$w_x = 0.75(41.8) = 31.4 \text{ lb/ft} < 67.0 \text{ lb/ft}$$

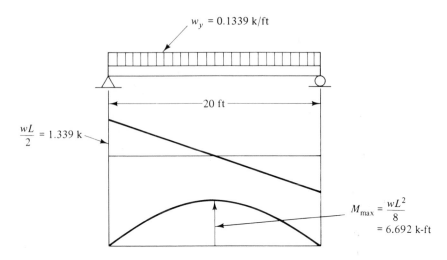

Figure 5-25   Example 5-22.

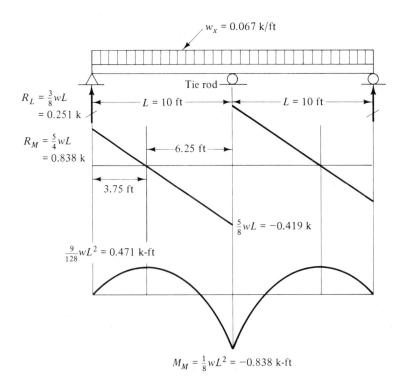

Figure 5-26   Example 5-22.

Both components of the load for case 2 are less than for case 1; therefore, consider case 1 only. The loading, supports, and shear and moment diagrams for the x–z plane are shown in Fig. 5-25 and for the y–z plane in Fig. 5-26. The shear forces and bending moments of Fig. 5-26 were calculated by superposition from formulas given on p. 2-124 of the AISC *Manual*, number 29. Assume that $S_x/S_y = 6.0$. From Eq. (5-21) the required section modulus

$$S_x \geq \frac{M_y + (S_x/S_y)M_y}{F_b} = \frac{[6.69 + 6(0.838)](12)}{22}$$

$$= 6.39 \text{ in.}^3$$

(a) For a depth $\frac{1}{30}$ span $= 20(12)/30 = 8$ in. From the allowable stress design selection table, try a W8 × 10: $S_x = 7.81$ in.^{3}, $S_y = 1.06$ in.^{3}. From Eq. (5-21),

$$S_x = \frac{[6.69 + (7.81/1.06)(0.838)](12)}{22}$$

$$= 7.02 \text{ in.}^3 < S_x = 7.81 \text{ in.}^3$$

Use a W8 × 10.

(b) For a depth $\frac{1}{24}$ span $= 20(12)/24 = 10$ in. From the allowable stress design selection table, try M10 × 9. The section is unsatisfactory. The second trial, a W10 × 12, is satisfactory. Use a W10 × 12.

5-16 OPEN-WEB STEEL JOISTS

Open-web steel joists (Fig. 5-27) are shop-fabricated lightweight trusses. They are convenient and very economical when used as floor or roof beams in buildings. The joists are relatively flexible members and depend on the floor or roof deck and bracing, called bridging, that spans between and transverse

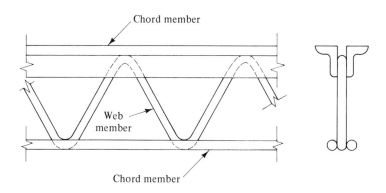

Figure 5-27 Open-web steel joist.

to the joists to provide lateral supports. Joists with a clear span up to 144 ft are available and the clear span-to-depth ratio normally does not exceed 24 at midspan.

The joists are classified as the H-series, the Longspan LH-series, and the Deep Longspan DHL-series. Specifications for each series have been established by the Steel Joist Institute (SJI). Membership in the SJI is open to manufacturers of joists that conform to the specifications and load tables.

When joists are desired the designer makes the selection from tables available in joist manufacturers' catalogs. The tables give the safe uniform load the joists can support, the safe end reactions, and the uniform load the joists can support if the deflection does not exceed $\frac{1}{360}$ of the span.

PROBLEMS

Use the AISC *Specifications* for the following problems. In design problems determine the lightest section.

5-1. A simply supported beam with a span of 12 ft has a 5-k load 4 ft from one end of the beam. Select a W section made of A36 steel that can safely support the load. Assume full lateral support.

5-2. Repeat Prob. 5-1 using A572 Grade 60 steel.

5-3. A simply supported beam with a span of 18 ft has a uniform load of 5 k/ft over the entire span. Select a W section of A36 steel that can safely support the load. Assume full lateral support.

5-4. Repeat Prob. 5-3 using A572 Grade 42 steel.

5-5. A simply supported beam with a span of 24 ft has a uniform load of 4 k/ft over the entire span. In addition, it has a 25-k load 8 ft from one end of the beam and a 20-k load 8 ft from the other end of the beam. Select a W section of A36 steel that can safely support the loads. Assume full lateral support.

5-6. Repeat Prob. 5-5 using A572 Grade 50 steel.

5-7. A simply supported beam has a span and loads as shown. Select a W section of

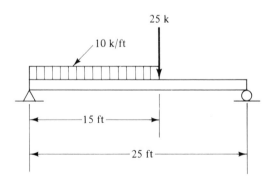

Prob. 5-7

A572 Grade 50 steel that can safely support the loads. Assume full lateral supports.

5-8. Repeat Prob. 5-7 using A572 Grade 60 steel.

5-9. A cantilever beam has loads and span as shown. Select a W section of A36 steel that can safely support the loads. Assume full lateral support for the compression flange.

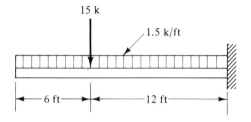

Prob. 5-9

5-10. Repeat Prob. 5-9 using A572 Grade 42 steel.

5-11. A beam with an overhanging end has loads and span as shown. Select a W section of A572 Grade 50 steel that can support the loads. Assume full lateral supports for the compression flange.

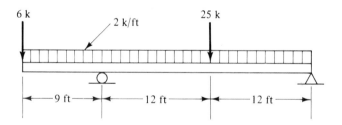

Prob. 5-11

5-12. Repeat Prob. 5-11 using A572 Grade 65 steel.

5-13. Beams and girders are used to support a reinforced concrete floor as shown in the figure. Beams, girders, and floor continue on all sides and the floor provides full lateral support. Assume that all beams and girders are simply supported. Reinforced concrete weighs 150 lb/ft^3. Use A36 steel to design beam B_1 and girder G_1 if the live load, floor thickness, and spans L_1 and L_2 are as follows:
   (a) LL = 70 psf, $t = 4$ in., $L_1 = 20$ ft, $L_2 = 21$ ft.
   (b) LL = 110 psf, $t = 4$ in., $L_1 = 19$ ft, $L_2 = 20$ ft.
   (c) LL = 125 psf, $t = 6$ in., $L_1 = 21$ ft, $L_2 = 24$ ft.
   (d) LL = 150 psf, $t = 6$ in., $L_1 = 23$ ft, $L_2 = 26$ ft.

5-14. Repeat Prob. 5-13 using A572 Grade 50 steel.

5-15. Repeat Prob. 5-13 assuming that the deflection limitations given by AISC Specification Commentary 1.13.1 apply.

5-16. Repeat Prob. 5-13 assuming that the floor beams are subject to shock loads and that provisions of AISC Specification Commentary 1.13.2 apply.

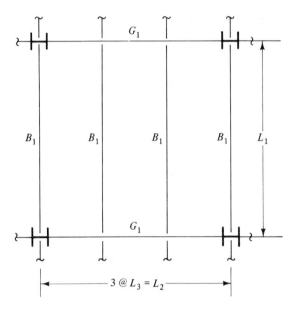

Prob. 5-13

5-17. A small building has steel framing as shown in the figure. A concrete floor provides full lateral support. Assume that beams and girders are simply supported. Use A26 steel. Reinforced concrete weighs 150 lb/ft³. Design each beam and girder if the live load, floor thickness, and spans L_1 and L_2 are as follows:

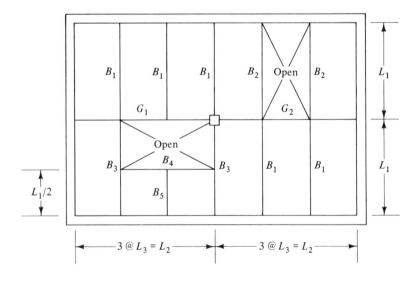

Prob. 5-17

(a) LL =   80 psf, $t = 4$ in., $L_1 = 21$ ft, $L_2 = 21$ ft.
(b) LL = 110 psf, $t = 4$ in., $L_1 = 20$ ft, $L_2 = 20$ ft.
(c) LL = 130 psf, $t = 6$ in., $L_1 = 25$ ft, $L_2 = 24$ ft.
(d) LL = 150 psf, $t = 6$ in., $L_1 = 26$ ft, $L_2 = 26$ ft.

5-18. Repeat Prob. 5-17 using A572 Grade 50 steel.

5-19. Lateral supports are provided at the ends of a simply supported beam made of A36 steel. Determine the allowable bending stress and allowable bending moment for the beam if it has a span of L_1, L_2, or L_3. Beam and spans are as follows:
(a) W27 × 102, $L_1 = 8$ ft, $L_2 = 10$ ft, $L_3 = 15$ ft.
(b) W24 × 84, $L_1 = 5$ ft, $L_2 = 12$ ft, $L_3 = 16$ ft.
(c) W21 × 44, $L_1 = 4$ ft, $L_2 = 6.9$ ft, $L_3 = 12$ ft.
(d) W18 × 76, $L_1 = 9$ ft, $L_2 = 13$ ft, $L_3 = 20$ ft.

5-20. Repeat Prob. 5-19 using A572 Grade 50 steel.

5-21. Lateral supports are provided at the ends of a simply supported beam made of A36 steel. A concentrated load acts at the middle of the span. Determine the allowable concentrated load that the beam can support if the beam section and span are as follows:
(a) W14 ×   43, $L = 20$ ft.
(b) W21 ×   73, $L = 22$ ft.
(c) W30 × 124, $L = 25$ ft.
(d) W36 × 182, $L = 23$ ft.

5-22. Repeat Prob. 5-21 using A527 Grade 50 steel.

5-23. Design a bearing plate for the following beams with end reaction, limit on length of bearing plate, and allowable bearing stress in the masonry under the plate.
(a) W10 ×   68, $R = 35$ k, $N = 10$ in., $F_p = 0.250$ ksi.
(b) W14 × 132, $R = 68$ k, $N = 12$ in., $F_p = 0.300$ ksi.
(c) W21 × 132, $R = 95$ k, $N = 12$ in., $F_p = 0.350$ ksi.
(d) W18 × 60, $R = 60$ k, $N = 10$ in., $F_p = 0.400$ ksi.
Use A36 steel, the width and length of the plate in inches, and the thickness of the plate in multiples of $\frac{1}{8}$ in.

5-24. Repeat Prob. 5-23 using A572 Grade 50 steel.

5-25. Select the lightest available section for a simply supported beam with span and uniform load over the entire span as follows:
(a) $L = 26$ ft, $w = 5.1$ k/ft.
(b) $L = 30$ ft, $w = 2.4$ k/ft.
(c) $L = 32$ ft, $w = 5.0$ k/ft.
(d) $L = 34$ ft, $w = 4.9$ k/ft.
Use A36 steel and assume that the beam is laterally supported at the ends only.

5-26. Repeat Prob. 5-25 assuming lateral supports at the ends and middle of the beam.

5-27. Select the lightest available section for a simply supported beam with a span and concentrated load located at a distance a from the left end of the beam as follows:
(a) $L = 34$ ft, $P = 80$ k, $a = 18$ ft.
(b) $L = 26$ ft, $P = 78$ k, $a = 10$ ft.
(c) $L = 30$ ft, $P = 40$ k, $a = 12$ ft.
(d) $L = 32$ ft, $P = 85$ k, $a = 12$ ft.
Use A36 steel and assume that the beam is laterally supported at the ends only.

5-28. Repeat Prob. 5-27 assuming lateral supports at the ends of the beam and at the concentrated load.

5-29. Select the lightest available section for a simply supported beam with span and loads as shown. Use A36 steel. Lateral supports are as follows:
  (a) Full lateral supports.
  (b) Lateral supports at the ends only.
  (c) Lateral supports at the ends and middle of the beam.
  (d) Lateral supports at the ends and third points (concentrated loads) of the beam.

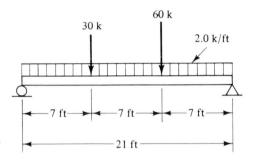

Prob. 5-29

5-30. Repeat Prob. 5-29 using A588 steel with $F_y = 50$ ksi.

5-31–5-35. Select the most economical available section for the beam with the lateral supports as indicated in the figure.
  (a) Use A36 steel.
  (b) Use A572 Grade 50 steel.

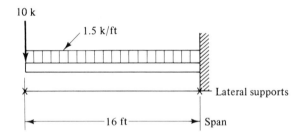

Prob. 5-31

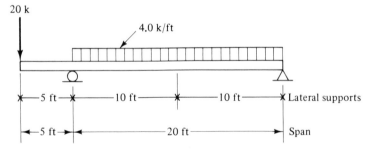

Prob. 5-32

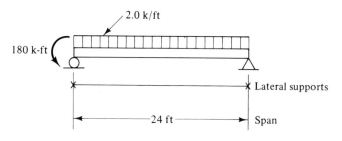

Prob. 5-33

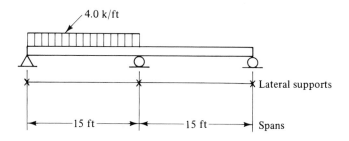

Prob. 5-34

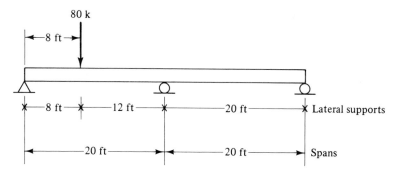

Prob. 5-35

6 BENDING AND AXIAL FORCE MEMBERS
Beam-Columns

6-1 INTRODUCTION

In Chapters 3 through 5, we considered the design of tension members, axially loaded compression members or columns, and bending members or beams. In this chapter we are concerned principally with the beam-column—a member in which the stresses caused by compressive axial loads and bending moments are both important. However, in Sec. 6-8 we also consider members under combined axial tension and bending.

When a member under combined axial tension and bending fails, the failure is usually due to yielding. A member under combined axial compression and bending may also fail due to yielding. More likely, failure will occur because of a number of possible unstable conditions. These include buckling in the plane of bending, buckling out of the plane of bending, and lateral buckling. Because of the many possible types of failure, no simple design procedure is possible.

In this chapter beam-columns are designed by semiempirical formulas based on allowables stress (AISC Specification 1.6.1). The following discussion is intended to provide some of the reasoning behind the use of these formulas.

6-2 ADDITIONAL BENDING DUE TO BENDING DEFLECTION

When a member is acted on by loads that produce bending, stresses and deflections occur in the member. From strength of materials, when equal

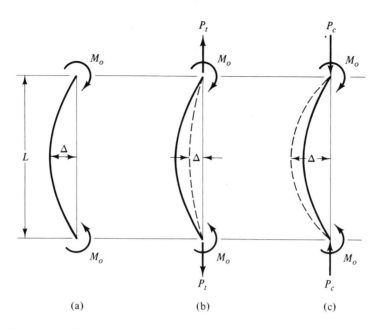

Figure 6-1 Effect of secondary moment $P\Delta$ on tension members and columns.

bending couples M_o are applied on the ends of the member as shown in Fig. 6-1(a), the maximum bending stress

$$f_b = +\frac{Mc}{I} = +\frac{M_o c}{I}$$ (a)

and the deflection at the middle of the member

$$\Delta = \frac{ML^2}{8EI} = \frac{M_o L^2}{8EI}$$ (b)

The application of a tensile load P_t as shown in Fig. 6-1(b) *reduces* the bending moment at the middle of the member by the secondary moment $P_t\Delta$. The maximum bending stress becomes

$$f_b = +\frac{Mc}{I} = +\frac{(M_o - P_t\Delta)c}{I}$$ (c)

and the *approximate* deflection at the middle of the member is given by

$$\Delta = \frac{ML^2}{8EI} = \frac{M_o L^2}{8EI} - \frac{P_t L^2}{8EI}\Delta$$ (d)

Solving for the bending stress from Eq. (c) requires the solution of Eq. (d) for Δ. The bending stress and deflection are both contracted or reduced when an axial tensile load acts with a bending moment on a member. Therefore, it is conservative and safe to neglect the secondary moment $P_t\Delta$ in problems involving bending and a tensile load.

When a compressive load P_c acts on the members as shown in Fig. 6-1(c), the bending moment *increases* at the middle of the beam by the secondary moment $P_c\Delta$. The maximum bending stress becomes

$$f_b = +\frac{Mc}{I} = +\frac{(M_0 + P_c\Delta)c}{I} \tag{e}$$

and the *approximate* deflection at the middle of the member is given by

$$\Delta = \frac{ML^2}{8EI} = \frac{M_0L^2}{8EI} + \frac{P_cL^2}{8EI}\Delta \tag{f}$$

When an axial compressive load acts together with a bending moment, both the bending stress and deflection are amplified or increased. As P_c approaches $8EI/L^2$, the last term of Eq. (f) approaches Δ. In that case, Δ approaches ∞ and buckling failure occurs.

There is a fundamental difference between tensile and compressive loads. An increase in the tensile load contracts or reduces the deflection and the effect of the secondary moment, while an increase in the compressive load amplifies or increases the deflection and the effect of the secondary moment. For critical values of the compressive load, buckling will occur. The effect of the secondary moment $P_c\Delta$ is provided for by the *amplification factor* in the AISC formulas for beam-columns.

6-3 COMBINED AXIAL COMPRESSION AND BENDING

To derive a basic interaction formula, consider a short compression member acted on by an axial compressive force and a bending couple. Neglecting the effect of deflection, the stresses in the member can be found by superposition as shown in Fig. 6-2. The maximum compressive stress by superposition is given by

$$f = \frac{P}{A} + \frac{Mc}{I} \tag{a}$$

Expressing the moment of inertia in terms of the radius of gyration $I = Ar^2$, we have

$$f = \frac{P}{A} + \frac{Mc}{Ar^2} \tag{b}$$

If $M = 0$, the required area A for the axial stress only is given by

$$A_a = \frac{P}{F_a} \tag{c}$$

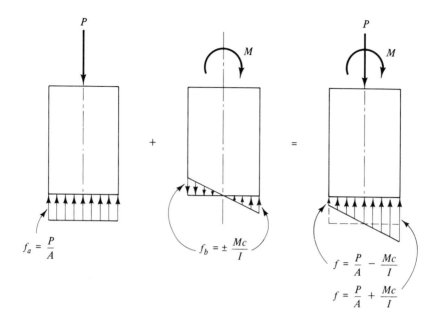

Figure 6-2   Superposition of axial and bending stresses.

where F_a is the allowable axial stress. If $P = 0$, the required area A_b for bending stress only is given by

$$A_b = \frac{Mc}{F_b r^2} \tag{d}$$

where F_b is the allowable bending stress. If the member is subject to both axial and bending loads, the total area required is the sum of the areas given by Eqs. (c) and (d). Thus

$$\frac{P}{F_a} + \frac{Mc}{F_b r^2} = A \tag{e}$$

Dividing both sides of Eq. (e) by the area A yields

$$\frac{P}{F_a A} + \frac{Mc}{F_b A r^2} = 1$$

Since $f_a = P/A$ and $f_b = Mc/I = Mc/Ar^2$, we have

$$\frac{f_a}{F_a} + \frac{f_b}{F_b} \leq 1 \tag{6-1}$$

Equation (6-1) is known as an *interaction formula.* A graph of f_a/F_a versus f_b/F_b takes the form of a straight line (Fig. 6-3). The formula may be thought of as a fractional formula. If, for example, 0.6 of the allowable axial stress is

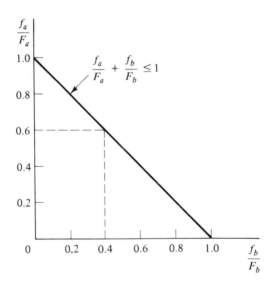

Figure 6-3   Graph of interaction formula.

used by f_a, then only 0.4 of the allowable bending stress may be used by f_b. The "less than" symbol ($<$) in Eq. (6-1) is used because fractional amounts adding up to less than 1.0 would also be safe.

For bending about both the x and y axes of the cross section, Eq. (6-1) can be extended to include both bending axes; thus

$$\frac{f_a}{F_a} + \frac{f_{bx}}{F_{bx}} + \frac{f_{by}}{F_{by}} \leq 1 \qquad (6\text{-}2)$$

6-4 AISC SPECIFICATIONS

Interaction equations similar to Eq. (6-2) have been used for many years. The present AISC Specification 1.6.1 is more conservative. To account for the additional secondary bending moment $P_c\Delta$ due to member deflection, the bending stress is multiplied by an *amplification factor*:

$$\frac{1}{1 - f_a/F'_e}$$

where f_a is the actual axial stress produced by the applied load and F'_e is the Euler buckling stress [Eq. (4-3)] divided by $\frac{23}{12}$, the AISC factor of safety for a long column with KL/r greater than C_c. The amplification factor is always greater than 1. For some conditions—depending on column location, end conditions, and type of load—the amplification factor is too conservative and the bending stress is multiplied by a reduction coefficient C_m.

According to AISC Specification 1.6.1, for a beam-column to be satisfactory for a given loading, the following equations must be satisfied:

$$\frac{f_a}{F_a} + \frac{C_{mx}f_{bx}}{(1 - f_a/F'_{ex})F_{bx}} + \frac{C_{my}f_{by}}{(1 - f_a/F'_{ey})F_{by}} \leq 1.0 \qquad (6\text{-}3)$$

[AISC Formula (1.6-1a)]

$$\frac{f_a}{0.6F_y} + \frac{f_{bx}}{F_{bx}} + \frac{f_{by}}{F_{by}} \leq 1.0 \qquad (6\text{-}4)$$

[AISC Formula (1.6-1b)]

When $f_a/F_a \leq 0.15$, Eq. (6-5) may be used in place of Eqs. (6-3) and (6-4):

$$\frac{f_a}{F_a} + \frac{f_{bx}}{F_{bx}} + \frac{f_{by}}{F_{by}} \leq 1.0 \qquad (6\text{-}5)$$

[AISC Formula (1.6-2)]

The subscripts x and y in these equations indicate the axis of bending about which a particular stress or design property applies, and

F_a = allowable axial compressive stress for axial force alone, ksi

F_b = allowable bending compressive stress for bending moments alone, ksi

$$F'_e = \frac{12\pi^2 E}{23(KL_b/r_b)^2} \quad (\text{AISC } \textit{Specifications, Appendix A, Table 9})$$

where L_b is the actual unbraced length, r_b is the corresponding radius of gyration, and K is the effective length factor, *all in the plane of bending.*

f_a = actual axial compressive stress, ksi

f_b = actual maximum compressive bending stress, ksi

C_m = reduction factor, determined by the following categories

1. *Sidesway permitted*: $C_m = 0.85$, compression members in frames subject to joint translation or sidesway. Sidesway is defined as relative motion of one end of a column with respect to the other end.

2. *No sidesway or transverse loads between ends of the beam:*

$$C_m = 0.6 - 0.4\frac{M_1}{M_2} \qquad \text{but not less than } 0.4$$

where M_1/M_2 is the ratio of the smaller to the larger moment at the ends of that portion of the member unbraced in the plane of bending under consideration. M_1/M_2 is positive when the member is bent in reverse curvature and negative when bent in single curvature.

3. *No sidesway with transverse loads between end supports:*
   a. For members with restrained ends $C_m = 0.85$.
   b. For members with unrestrained ends $C_m = 1.0$.
   In place of these values of C_m the reduction factor may be determined
   for various end conditions and loads from the equation

   $$C_m = 1.0 + \psi \frac{f_a}{F'_e}$$

where f_a = actual axial stress
       F'_e = Euler stress as previously defined with a factor of safety of
       $\frac{23}{12}$
       ψ = factor from Fig. 6-4; the factor depends on end restraints
       and transverse loads

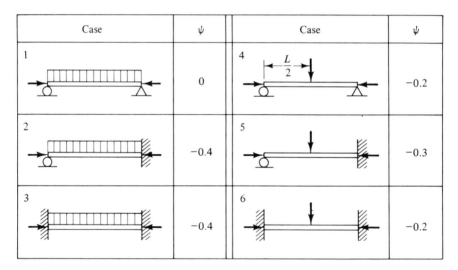

Case	ψ	Case	ψ
1	0	4	−0.2
2	−0.4	5	−0.3
3	−0.4	6	−0.2

Figure 6-4 ψ Factor for various cases of end restraints and transverse
loads $(C_m = 1 + \psi f_a / F'_e)$. (Adapted from Table C1.6.1 of the AISC
Specification Commentary.)

6-5 ANALYSIS OF BEAM-COLUMNS (AISC METHOD)

The equations that must be satisfied for beam-columns [Eqs. (6-3) and (6-4)]
or (6-5)] are fairly complicated. They require calculations that combine the
techniques and methods for both axially loaded columns and for beams. Hence
it will be helpful to separate the column action from the beam action in our
calculations. In the following we consider examples that will help clarify the
problem areas.

Example 6-1

A W16 × 77 section of A36 steel is used as a beam-column with loading as shown in Fig. 6-5. The ends of the member are pinned and braced against sidesway. Is the member satisfactory?

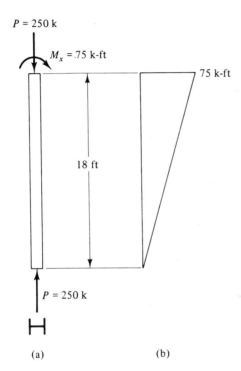

P = 250 k

$M_x = 75$ k-ft

75 k-ft

18 ft

P = 250 k

(a)   (b)   **Figure 6-5** Example 6-1.

Solution. Properties of the W16 × 77: $A = 22.6$ in.^{2}, $S_x = 134$ in.^{3}, $r_x = 7.00$ in., $r_y = 2.47$ in., $r_T = 2.77$ in., $d/A_f = 2.11$ in.$^{-1}$.

Column action. The controlling slenderness ratios

$$\frac{K_y L_y}{r_y} = \frac{1.0(18)(12)}{2.47} = 87.5$$

From Table 3-36 of the AISC *Specifications*, Appendix A, the allowable axial stress

$$F_a = 14.51 \text{ ksi}$$

and the actual axial stress

$$f_a = \frac{P}{A} = \frac{250}{22.6} = 11.06 \text{ ksi}$$

Beam action. The moment diagram is shown in Fig. 6-5(b). The distance between lateral supports $L_b = 18$ ft is greater than $L_u = 15.5$ ft and the section is compact. Therefore, the allowable bending stress is found from Eqs. (5-16) or (5-17) and (5-18) [AISC Formulas (1.5-6a) or (1.5-6b) and (1.5-7)]. Values of C_b are used in these formulas for the allowable bending stress F_b as follows:

1. For use in Eq. (6-3) [AISC Formula (1.6-1a)] when the ends of the member are braced against sidesway,

$$C_b = 1.0$$

2. For use in Eq. (6-4) [AISC Formula (1.6-1b)],

$$C_b = 1.75 + 1.05\left(\frac{M_1}{M_2}\right) + \left(\frac{M_1}{M_2}\right)^2$$

A bending factor of $C_b = 1.0$ is used in Eq. (6-3) because the reduction factor C_m in that formula takes care of the moment gradient (nonuniform bending moment). The slenderness ratio

$$\frac{L}{r_T} = \frac{18(12)}{2.77} = 78.0 < \frac{L}{r_T} = 119$$

Therefore, from Eq. (5-16) [AISC Formula (1.5-6a)] with $F_y = 36$ ksi,

$$F_{b1} = 24.0 - \frac{(78.0)^2}{1181} = 18.85 \text{ ksi}$$

and Eq. (5-18) [AISC Formula (1.5-7)],

$$F_{b2} = \frac{12(10)^3}{18(12)(2.11)} = 26.3 \text{ ksi} > F_b = 22 \text{ ksi}$$

Use $F_b = 22$ ksi. With $C_b > 1.0$ the allowable bending stress is unchanged. The actual bending stress

$$f_b = \frac{M}{S} = \frac{75(12)}{134} = 6.72 \text{ ksi}$$

The reduction factor from category 2,

$$C_m = 0.6 - 0.4\frac{M_1}{M_2} = 0.6 - 0.4\left(\frac{0}{115}\right)$$

$$= 0.6$$

The slenderness ratio in the plane of bending

$$\frac{K_b L_b}{r_b} = \frac{K_x L_x}{r_x} = \frac{18(12)}{7.00} = 30.9$$

and the stress from Table 9 of the AISC *Specifications*, Appendix A,

$$F'_e = 156.9 \text{ ksi}$$

From Eqs. (6-3) and (6-4) [AISC Formulas (1.6-1a) and (1.6-1b)],

$$\frac{f_a}{F_a} + \frac{C_m f_b}{(1 - f_a/F'_e)F_b} = \frac{11.06}{14.51} + \frac{0.6(6.72)}{(1 - 11.06/156.9)22}$$

$$= 0.96 < 1.0 \qquad \text{OK}$$

$$\frac{f_a}{0.6F_y} + \frac{f_b}{F_b} = \frac{11.06}{22} + \frac{6.72}{22} = 0.81 < 1.0 \qquad \text{OK}$$

The W16 × 77 is satisfactory.

Example 6-2

A steel section formed from two L6 × 6 × $\frac{7}{8}$ separated by $\frac{3}{8}$ in. and made of A36 steel is used as a continuous top chord member of the roof truss shown in Fig. 6-6(a). The roofing is directly supported by the chord and furnishes lateral support for the member. The loads acting on member *FG* are shown in Fig. 6-6(b). Is the member satisfactory?

Solution. The top chord of the truss is assumed to be pinned at *F* and *I* and continuous over supports *G* and *H*. From Beam Diagrams and Formulas, number 36, of the AISC *Manual*, the moment diagram is constructed in Fig. 6-6(c). Properties of two L6 × 6 × $\frac{7}{8}$ separated by $\frac{3}{8}$ in.: $A = 19.5$ in.^{2}, $S_x = 15.3$ in., $r_x = 1.81$ in., $r_y = 2.70$ in., and Q_s is not listed; therefore, the angles comply with AISC Specification 1.9.1.2 and may be considered fully effective.

Column action. The controlling slenderness ratio

$$\frac{K_x L_x}{r_x} = \frac{1.0(14)(12)}{1.81} = 92.8$$

From Table 3-36 of the AISC *Specifications*, Appendix A, the allowable axial stress

$$F_a = 13.84 \text{ ksi}$$

and the actual axial stress

$$f_a = \frac{P}{A} = \frac{105}{19.5} = 5.38 \text{ ksi}$$

Beam action. The bending stress at the end of the member

$$f_{b2} = \frac{M_2}{S} = \frac{14.7(12)}{15.3} = 11.53 \text{ ksi}$$

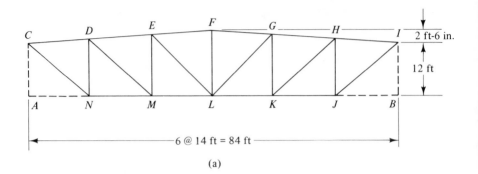

(a)

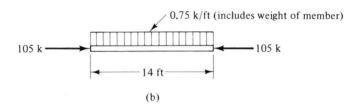

(b)

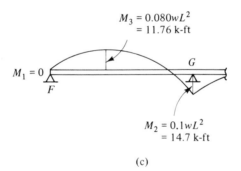

(c)

Figure 6-6 Example 6-2.

and near the middle of the member

$$f_{b3} = \frac{M_3}{S} = \frac{11.76(12)}{15.3} = 9.22 \text{ ksi}$$

The slenderness ratio in the plane of bending

$$\frac{K_b L_b}{r_b} = \frac{K_x L_x}{r_x} = \frac{14(12)}{1.81} = 92.8$$

and the allowable stress from Table 9 of the AISC *Specifications*, Appendix A,

$$F'_e = 17.34 \text{ ksi}$$

The angles have continuous lateral supports; therefore, $F_b = 22$ ksi. The reduction factor from category 3, case 2,

$$C_m = 1 - \psi \frac{f_a}{F_e'} = 1 - 0.4 \left(\frac{5.38}{17.34} \right) = 0.875$$

From Eqs. (6-3) and (6-4) [AISC Formulas (1.6-1a) and (1.6-1b)],

$$\frac{f_a}{F_a} + \frac{C_m f_{b3}}{(1 - f_a/F_e')F_b} = \frac{5.38}{13.86} + \frac{0.875(9.22)}{(1 - 5.38/17.34)22}$$

$$= 0.92 < 1.0 \qquad \text{OK}$$

$$\frac{f_a}{0.6F_y} + \frac{f_{b2}}{F_b} = \frac{5.38}{22} + \frac{11.53}{22} = 0.77 < 1.0 \qquad \text{OK}$$

The two L6 × 6 × $\frac{7}{8}$ separated by $\frac{3}{8}$ in. are satisfactory.

Example 6-3

A W10 × 33 section of A36 steel is used as a beam-column as shown in Fig. 6-7(a). The member is laterally braced at its ends and at midheight in the weak direction (perpendicular to the web). Is the member satisfactory?

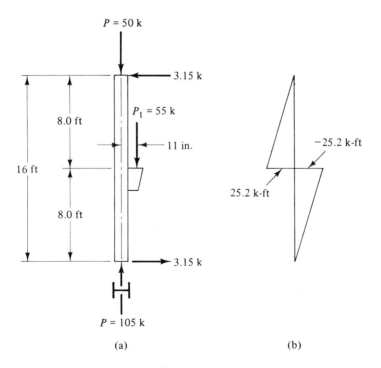

(a)   (b)

Figure 6-7   Example 6-3.

Solution. Properties of a W10 × 33: $A = 9.71$ in.^{2}, $S_x = 35.0$ in.^{3}, $r_x = 4.19$ in., $r_y = 1.94$ in., $L_c = 8.4$ ft, $L_u = 16.5$ ft.

Column action. The slenderness ratios

$$\frac{K_x L_x}{r_x} = \frac{1.0(16)(12)}{4.19} = 45.8 \qquad \frac{K_y L_y}{r_y} = \frac{1.0(8)(12)}{1.94} = 49.5$$

For a controlling slenderness ratio of 49.5, the stress from Table 3-36 of the AISC *Specifications*, Appendix A, is $F_a = 18.40$ ksi. The actual axial stress $f_a = P/A = 105/9.71 = 10.81$ ksi.

Beam action. The reactions at the top and bottom of the member were calculated and the moment diagram drawn as illustrated in Fig. 6-7(b). The maximum bending stress

$$f_b = \frac{M}{S} = \frac{25.2(12)}{35} = 8.64 \text{ ksi}$$

The distance between lateral supports $L_b = 8$ ft $< L_c = 8.4$ ft. Therefore, $F_b = 0.66 F_y = 24$ ksi.

In the plane of bending

$$\frac{K_b L_b}{r_b} = \frac{K_x L_x}{r_x} = 45.8$$

From Table 9 of the AISC *Specifications*, Appendix A, the stress $F'_e = 71.1$ ksi. The loading and moment diagram does not fit any of the categories; therefore, conservatively we use a reduction factor $C_m = 1.0$. From Eqs. (6-3) and (6-4) [AISC Formulas (1.6-1a) and (1.6-1b)],

$$\frac{10.81}{18.14} + \frac{1.0(8.64)}{(1 - 10.81/71.1)24} = 1.01 > 1.0 \qquad \text{NG}$$

$$\frac{10.81}{22} + \frac{8.64}{24} = 0.85 < 1.0 \qquad \text{OK}$$

The W10 × 33 is probably satisfactory. Equation (6.3) exceeds 1.0 by approximately 1.0 percent.

6-6 DESIGN OF BEAM-COLUMNS (AISC METHOD)

For convenience in design, Eqs. (6-3), (6-4), and (6-5) [AISC Formulas (1.6-1a), (1.6-1b), and (1.6-7)] can be written in a modified form that in effect converts the bending moments into equivalent axial loads. With bending about the x and y axes the three modified equations are as follows:

$$P + P'_x + P'_y = P + \left[B_x M_x C_{mx} \left(\frac{F_a}{F_{bx}} \right) \frac{a_x}{a_x - P(KL)^2} \right]$$
$$+ \left[B_y M_y C_{my} \left(\frac{F_a}{F_{by}} \right) \frac{b_x}{a_y - P(kL)^2} \right] \qquad (6\text{-}7)$$

[Modified AISC Formula (1.6-1a)]

$$P + P'_x + P'_y = P \left(\frac{F_a}{0.6F_y} \right) + \left[B_x M_x \left(\frac{F_a}{F_{bx}} \right) \right] + \left[B_y M_y \left(\frac{F_a}{F_{by}} \right) \right] \qquad (6\text{-}8)$$

[Modified AISC Formula (1.6-1b)]

When $f_a / F_a = 0.15$,

$$P + P'_x + P'_y \doteq P + \left[B_x M_x \left(\frac{F_a}{F_{bx}} \right) \right] + \left[B_y M_y \left(\frac{F_a}{F_{by}} \right) \right] \qquad (6\text{-}9)$$

[Modified AISC Formula (1.6-7)]

where B = bending factor A/S
$\quad\; a = 149{,}000 Ar^2$ for the axis of bending
$\quad\; K$ = the effective length factor
$\quad\; L$ = actual unbraced length in the plane of bending

The sum of the load terms on the left-hand side of the modified equations can be thought of as an equivalent axial load. With the equivalent axial load and the effective length a member can be selected from the column load tables in the AISC *Manual.* Values of B_x, B_y, a_x, and a_y are listed at the bottom of the column load tables.

Methods for Selecting Trial Column Sections

Equivalent axial loads can often be determined by simple empirical equations for various structural shapes as follow:

W, M, or S shapes:

$$P_{eq} = P + 25.2 \frac{M_x}{d} + 72.0 \frac{M_y}{d} \qquad (6\text{-}10)$$

Two equal-leg angles (back to back):

$$P_{eq} = P + 72.0 \frac{M_x}{L} \qquad (6\text{-}11)$$

WT shapes:

$$P_{eq} = P + 127 \frac{M_x}{d} \qquad (6\text{-}12)$$

where

P_{eq} = equivalent axial load, k

M_x, M_y = maximum bending moment about the x and y axes, k-ft

d = nominal depth of the section, in.

L = leg length for angle, in.

Another method for selecting an economical trial W, M, or S shape based on an equivalent axial load is given on p. 3-10 of the AISC *Manual.* Each method is illustrated in the following examples.

Example 6-4   *Beam-Column Selection*

Select the lightest W12 section for a column 20 ft long that supports an axial load $P = 180$ k and a bending moment at the top of $M_x = 90$ k-ft and the bottom of $M_x = 115$ k-ft (Fig. 6-8). The column is braced against sidesway and pinned at the top and bottom in the x-z plane and fixed at the top and bottom in the y-z plane and sidesway is permitted. Use A36 steel.

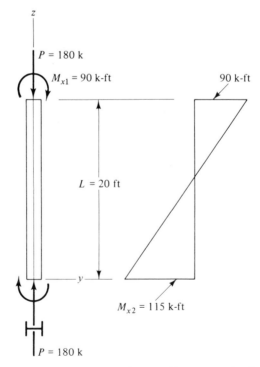

(y-z plane: Fixed top and bottom sidesway permitted)
(x-z plane: Pinned top and bottom sidesway prevented)

Figure 6-8   Example 6-4.

Solution. From Eq. (6-10) we make a tentative W12 selection:

$$P_{eq} = P + 25.2\frac{M_x}{d} = 180 + \frac{25.2(115)}{12} = 422\,k$$

From the column load tables of the AISC *Manual*, p. 3-24, select a W12 × 87.

From the method of the AISC *Manual*, p. 3-10, we make a second tentative selection. The reduction factor from category 2

$$C_m = 0.6 - 0.4\frac{M_1}{M_2} = 0.6 - 0.4\frac{90}{115} = 0.287$$

but not less than 0.4. Use $C_m = 0.4$. From Table B with $KL = 20$ ft and a W12 shape, $m = 2.0$. Therefore (see the note at the bottom of Table B),

$$m' = \frac{C_m m}{0.85} = \frac{0.4(2.0)}{0.85} = 0.941$$

and

$$P_{eq} = P + M_x m' = 180 + 115(0.941) = 288\,k$$

From the column load tables select a W12 × 65.

Two column sections fall between the W12 × 65 and W12 × 87. They are the W12 × 79 and W12 × 72.

Check of W12 × 72. From the column load tables:

$$L_c = 12.8\,ft < L_b = 20\,ft < L_u = 33.3\,ft$$

The section is compact; therefore, $F_b = 22$ ksi.

$$A = 21.1\,in.^2,\ P_{all} = 326\,k,\ r_x = 5.34\,in.$$
$$r_y = 3.04\,in.,\ B_x = 0.217,\ a_x = 88.6(10)^6$$

The slenderness ratios

$$\frac{K_x L_x}{r_x} = \frac{1.2(20)(12)}{5.34} = 53.9 \qquad \frac{K_y L_y}{r_y} = \frac{1.0(20)(12)}{3.04} = 78.9$$

For the controlling slenderness ratio of 78.9, the allowable axial stress $F_a = 15.48$ ksi. The actual axial stress $f_a = 180/21.1 = 8.53$ ksi. In the plane of bending

$$K_x L_x = 1.2(20)(12) = 288\,in.$$
$$P(K_x L_x)^2 = 180(288)^2 = 14.93(10)^6$$

From Eqs. (6-6) and (6-7) [Modified AISC Formulas (1.6-1a) and (1.6-1b)],

$$P + P'_x = 180 + \left[0.217(115)(12)(0.4)\left(\frac{15.48}{22}\right) \right.$$

$$\left. \times \left(\frac{88.6(10)^6}{88.6(10)^6 - 14.93(10)^6}\right) \right]$$

$$= 281.4 \text{ k} < P_{\text{all}} = 326 \text{ k} \qquad \text{OK}$$

$$P + P'_x = 180\left(\frac{15.48}{22}\right) + \left[0.217(115)(12)\left(\frac{15.48}{22}\right) \right]$$

$$= 337.4 \text{ k} > P_{\text{all}} = 326 \text{ k} \qquad \text{NG}$$

The W12 × 72 is not satisfactory. The next section is satisfactory. Use a W12 × 79.

Example 6-5   *Beam-Column Selection*

Select the lightest W14 section for a floor beam acting as a frame member that supports an axial load $P = 50$ k and a uniformly distributed load $w = 1.9$ k/ft, not including the weight of the beam as shown in Fig. 6-9. The beam is laterally braced at the ends only. Use A36 steel.

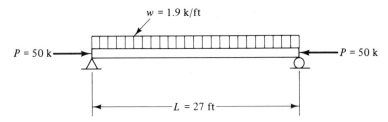

Figure 6-9   Example 6-5.

Solution. The maximum moment not including the weight of the beam $M_{\text{max}} = wL^2/8 = 1.9(27)^2/8 = 173$ k-ft. From Eq. (6-10),

$$P_{\text{eq}} = P + 25.2\frac{M_x}{d} = 50 + \frac{25.2(173)}{14} = 361 \text{ k}$$

From the column load tables, AISC *Manual*, p. 3-20, select a W14 × 90.

From the AISC *Manual* method, p. 3-10, for finding an equivalent load we make a second tentative selection. The reduction factor from category 3, case 1, is $C_m = 1.0$. From Table B with $KL = 27$ ft and a W14 shape, $m = 1.7$. Therefore (see the note at the bottom of Table B),

$$m' = \frac{C_m m}{0.85} = \frac{1.0(1.7)}{0.85} = 2.0$$

and

$$P_{eq} = P + M_x m' = 50 + 173(2.0) = 396 \text{ k}$$

From the column load table select a W14 × 90.

Check of W14 × 90. From the column load tables:

$$L_c = 15.3 \text{ ft} < L_b = 27 \text{ ft} < L_u = 34.0 \text{ ft}$$

The section is compact; therefore, $F_b = 22$ ksi.

$$A = 26.5 \text{ in.}^2, \ P_{all} = 384 \text{ k}, \ r_x = 6.14 \text{ in.}$$

$$r_y = 3.70 \text{ in.}, \ B_x = 0.185, \ a_x = 148.9(10)^6$$

The bending moment including the weight of the member

$$M_{max} = \frac{wL^2}{8} = \frac{1.990(27)^2}{8} = 181.3 \text{ k-ft}$$

The controlling slenderness ratio

$$\frac{K_y L_y}{r_y} = \frac{1.0(27)(12)}{3.70} = 87.6$$

and the allowable axial stress $F_a = 14.49$ ksi. The actual axial stress $f_a = P/A = 50/26.5 = 1.89$ ksi.

Since $f_a/F_a = 1.89/14.49 = 0.13 \leq 0.15$, use Eq. (6-5) [Modified AISC Formula (1.6-2)].

$$P + P'_x = 50 + \left[0.185(181.3)(12)\left(\frac{14.49}{22}\right) \right]$$

$$= 315 \text{ k} < P_{all} = 384 \text{ k} \qquad \text{OK}$$

Use a W14 × 90.

Example 6-6

Select the lightest W8 section of A36 steel for the member of a truss 16 ft long that supports an axial load $P = 175$ k and a transverse load $P_1 = 16$ k at midspan as shown in Fig. 6-10. Lateral bracing is provided at the ends of the member and at midspan perpendicular to the y axis of the cross section. The member is part of the continuous top chord of a truss. To calculate the bending moments, assume that the ends are fixed.

Solution.   For fixed ends the maximum moment not including the weight of the beam $M_{max} = P_1 L/8 = 16(16)/8 = 32$ k-ft. From Eq. (6-10), the equivalent load

$$P_{eq} = P + 25.2 \frac{M_x}{d} = 175 + \frac{25.2(32)}{8} = 276 \text{ k}$$

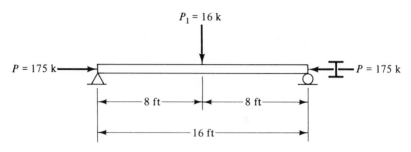

Figure 6-10   Example 6-6.

Select a W8 × 58 from the column load table of the AISC *Manual*, p. 3-28. The controlling slenderness ratio will be with respect to the x axis. Therefore, the effective slenderness ratio $K_x L_x / (r_x / r_y) = 1.0(16)/1.74 = 9.2$ ft.

From the AISC *Manual*, p. 3-10, we make a second trial selection. To be conservative, assume that the reduction coefficient from category 3 is $C_m = 1.0$. For $KL = 9.2$ ft and a W8 shape from Table B, $m = 3.0$. Therefore, $m' = C_m m / 0.85 = 1.0(3.0)/0.85 = 3.53$ and the equivalent load

$$P_{eq} = P + M_x m' = 175 + 32(3.53) = 288 \text{ k}$$

Select a W8 × 58 from the column load tables.

Check of W8 × 58. From the column load tables:

$$L_b = 8 \text{ ft} < L_c = 8.7 \text{ ft}$$

The section is compact and the allowable bending stress $F_b = 0.66 F_y = 24$ ksi.

$$A = 17.7 \text{ in.}^2, \qquad P_{all} = 310 \text{ k}, \qquad r_x = 3.65 \text{ in.}$$

$$r_y = 2.10 \text{ in.}, \qquad B_x = 0.329, \qquad a_x = 33.9(10)^6$$

The slenderness ratios

$$\frac{K_x L_x}{r_x} = \frac{1.0(16)(12)}{3.65} = 52.6 \qquad \frac{K_y L_y}{r_y} = \frac{1.0(8)(12)}{2.10} = 45.7$$

For a controlling slenderness ratio of 52.6, the allowable axial stress $F_a = 18.12$ ksi and the actual axial stress $f_a = P/A = 175/17.1 = 10.23$ ksi. Including the weight of the member, the maximum moments at the middle of the member

$$M_3 = \frac{PL}{8} + \frac{wL^2}{24} = \frac{16(16)}{8} + \frac{0.058(16)^2}{24} = 30.6 \text{ k-ft}$$

and the moments at the ends of the member

$$M_1 = M_2 = \frac{PL}{8} + \frac{wL^2}{12} = \frac{16(16)}{8} + \frac{0.058(16)^2}{12} = 33.2 \text{ k-ft}$$

In the plane of bending

$$KL = 1.0(16)(12) = 192$$

$$P(KL)^2 = 175(192)^2 = 6.45(10)^6$$

Since $f_a/F_a = 10.23/18.12 = 0.565 > 0.15$, we use Eqs. (6-7) and (6-8) [Modified AISC Formulas (1.6-1a) and (1.6-1b)]. That is,

$$P + P_x' = 175 + \left[0.329(30.6)(12)(1.0)\left(\frac{18.12}{24}\right) \right.$$

$$\left. \times \left(\frac{33.9(10)^6}{33.9(10)^6 - 6.45(10)^6}\right) \right]$$

$$= 288\ \text{k} < P_{\text{all}} = 310\ \text{k} \qquad \text{OK}$$

and

$$P + P_x' = 175\left(\frac{18.12}{22}\right) + \left[0.329(33.2)(12)\left(\frac{18.12}{24}\right) \right]$$

$$= 243\ \text{k} < P_{\text{all}} = 310\ \text{k} \qquad \text{OK}$$

Use a W8 × 58.

Example 6-7

Select the lightest W14 section made of A36 steel for the column shown in Fig. 6-11. The column is part of a braced frame. The load $P = 400\ \text{k}$ has an eccentricity $e = 12$ in. with respect to the x axis of the cross section. Assume that the column height $L = 20$ ft is the effective buckling length.

Solution.   From Eq. (6-10),

$$P_{\text{eq}} = P + 25.2\frac{M_x}{d} = 400 + \frac{25.2(400)}{14} = 1120\ \text{k}$$

From the column load tables in the AISC *Manual*, p. 3-18, we select a W14 × 233.

We make a second trial selection from the AISC *Manual*, p. 3-10. The reduction coefficient from category 2

$$C_m = 0.6 - 0.4\frac{M_1}{M_2} = 0.6 - 0.4\left(\frac{0}{400}\right) = 0.6$$

With $KL = 20$ ft and a W14 shape, $m = 1.7$ from Table B. Therefore,

$$m' = \frac{C_m m}{0.85} = \frac{0.6(1.7)}{0.85} = 1.2$$

and

$$P_{\text{eq}} = P + M_x m' = 400 + 400(1.2) = 880\ \text{k}$$

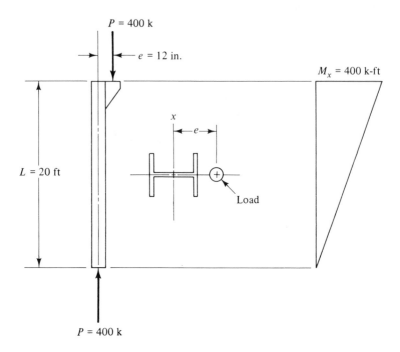

Figure 6-11   Example 6-7.

Select a W14 × 176 from the column load tables. Two sections fall between the W14 × 233 and the W14 × 176, the W14 × 193 that is unsatisfactory and the W14 × 211.

Check of the W14 × 211. From the column load tables:

$$L_c = 16.7 \text{ ft} < L_b = 20 \text{ ft} < L_u = 72.3 \text{ ft}$$

The section is compact and the allowable bending stress $F_b = 0.6F_y = 22$ ksi.

$$A = 62.0 \text{ in.}^2, \, P_{\text{all}} = 1089 \text{ k}, \, r_x = 6.55 \text{ in.}$$

$$r_y = 4.07 \text{ in.}, \, B_x = 0.183, \, a_x = 396(10)^6$$

The controlling slenderness ratio

$$\frac{K_y L_y}{r_x} = \frac{1.0(20)(12)}{4.07} = 59.0$$

The allowable axial stress $F_a = 17.53$ ksi and the actual axial stress including the weight of the member $f_a = P/A = [(400 + 0.211(20)]/62.0 = 6.52$ ksi. In the plane of bending

$$KL = 0.1(20)(12) = 240 \text{ in.}$$

$$P(KL)^2 = 404.2(240)^2 = 23.28(10)^6$$

Since $f_a/F_a = 6.52/17.53 = 0.372 > 0.15$, use Eqs. (6-7) and (6-8) [Modified AISC Formulas (1.6-1a) and (1.6-1b)]. Thus

$$P + P'_x = 404.2 + \left[0.183(400)(12)(0.6)\left(\frac{1753}{22}\right) \right.$$

$$\left. \times \left(\frac{396(10)^6}{396(10)^6 - 23.28(10)^6}\right) \right]$$

$$= 850.4 \text{ k} < P_{\text{all}} = 1087 \text{ k} \qquad \text{OK}$$

and

$$P + P'_x = 404.2\left(\frac{17.53}{22}\right) + \left[0.183(400)(12)\left(\frac{17.53}{22}\right) \right]$$

$$= 1022 \text{ k} < P_{\text{all}} = 1087 \text{ k} \qquad \text{OK}$$

Use a W14 × 211.

6-7 EFFECTIVE LENGTHS OF COLUMNS IN FRAMES

The concept of an effective length for columns was discussed in Chapter 4. Theoretical and recommended lengths for columns with various end conditions were given in Table 4-1. However, the table was developed for idealized end conditions that may be completely different from those in design. In reality a column is only part of a structure and the effective length for a particular column is a property of the structure.

A more general determination of the effective length of a column in continuous frames can be made by the alignment charts given in Fig. 6-12. The charts are based on the buckling equations for frames. They provide convenient solutions to equations that can otherwise be solved only by trial and error. Alignment charts for sidesway prevented and sidesway permitted are also given in the AISC *Manual*, p. 3-5, and a chart for sidesway permitted is given in the Commentary of AISC Specification, Sec. 1.8, Fig. C1.8.2.

The use of the charts shown in Fig. 6-12 requires the calculation of the stiffness ratios

$$G_A = \frac{\sum I_c/L_c \quad \text{(columns)}}{\sum I_b/L_b \quad \text{(beams)}} \tag{6-13}$$

and

$$G_B = \frac{\sum I_c/L_c \quad \text{(columns)}}{\sum I_b/L_b \quad \text{(beams)}} \tag{6-14}$$

where the subscripts A and B indicate the ends of the column being analyzed. (The values of G_A and G_B can be used interchangeably.) The $\sum$ indicates the

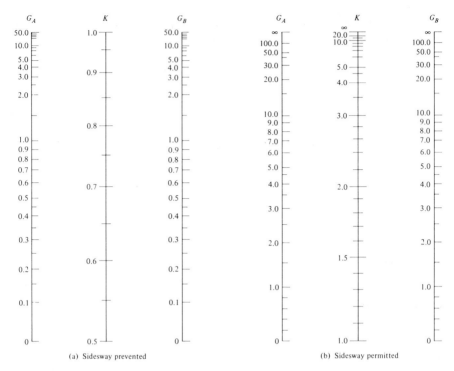

(a) Sidesway prevented   (b) Sidesway permitted

Figure 6-12 Chart for effective length of columns in continuous frames. (Adapted from AISC *Manual of Steel Construction*, 8th ed., Chicago, 1980. Courtesy of the American Institute of Steel Construction, Inc.)

summation of the ratios of the moment of inertia I to the unsupported length L for all the columns (c) or beams (b) rigidly connected to that joint and lying in the plane in which buckling of the column is being considered. The moment of inertia I is taken about an axis perpendicular to the plane of buckling.

Recommendations for Use of the Chart

1. When a column end is pinned, G is theoretically infinite. *Should use* $G = 10$.
2. When a column end is fixed, G is theoretically zero. *Should use* $G = 1.0$.
3. When a beam (girder) is rigidly attached to the column of interest, the beam stiffness I_b/L_b should be multiplied by an appropriate factor as follows:
   a. *Sidesway prevented* [Fig. 6-12(a)]
   (1) 1.5, for far end pinned
   (2) 2.0, for far end fixed against rotation

b. *Sidesway permitted* [Fig. 6-12(b)]

(1) 0.5, for far end pinned

(2) 0.67, for far end fixed against rotation

Having determined G_A and G_B for a column section, K is obtained by drawing a straight line between the two points on the G_A and G_B scales and reading the value from the K scale. For example, for sidesway prevented if $G_A = 0.346$ and $G_B = 0.479$, the value of $K = 0.66$.

For the assumptions on which the alignment charts are based, see the Commentary on the AISC Specification, p. 5-126 of the AISC *Manual.*

Inelastic Effects

When the KL/r ratio of a column is less than C_c, inelastic buckling should be considered. The AISC *Manual,* pp. 3-6 and 3-7, provides a direct method for considering inelastic buckling. The procedure is as follows:

1. Select a trial column size.
2. Calculate the actual axial stress, $f_a = P/A$.
3. From Table A, p. 3-7 of the AISC *Manual,* determine the reduction factor f_a/F'_e. (For values of f_a less than those shown in the table, the column is elastic and the reduction factor is 1.0.)
4. Calculate the inelastic stiffness ratios from the equation

$$G_{\text{inelastic}} = \frac{f_a}{F'_e} \, G_{\text{elastic}} \qquad (6\text{-}15)$$

for the top and bottom of the column. *No reduction in value should be made for a pinned end* $(G = 10.0)$ *or a fixed end* $(G = 1.0)$.

The method is illustrated in Example 6-11.

When sidesway is permitted, the effective length for a column is greater than its actual length. On the other hand, when sidesway is prevented, the effective length is less than its actual length.

Example 6-8

Using the alignment charts in Fig. 6-12, determine the effective length of each of the columns in the frame shown in Fig. 6-13 if

(a) Sidesway is prevented.

(b) Sidesway is permitted.

Solution.   The stiffnesses of each member, the G factors for each joint, and the effective length factors for each column were calculated and are tabulated in Tables (a), (b), and (c).

TABLE (a) for Example 6-8   Stiffness

Member	Section	I (in.^{4})	L (ft)	I/L (in.^{4}/ft)
AD, CF	W8 × 40	146	14	10.43
DG, FI	W8 × 40	146	12	12.17
BE	W8 × 64	272	14	19.43
EH	W8 × 64	272	12	22.67
DE, EF	W24 × 68	1830	28	65.36
GH, HI	W18 × 46	712	28	25.43

TABLE (b) for Example 6-8   G factors

Joint	$\dfrac{\sum I_c/L_c}{\sum I_b/L_b}$	G
A, C	See recommendations for use of chart, 1	10.0
B	See recommendations for use of chart, 2	1.0
D, F	$\dfrac{12.17 + 10.43}{65.36}$	0.346
E	$\dfrac{22.67 + 19.43}{2(65.36)}$	0.322
G, I	$\dfrac{12.17}{25.43}$	0.479
H	$\dfrac{22.67}{2(25.43)}$	0.446

TABLE (c) for Example 6-8   Effective length factor K from Fig. 6-12

Column	G_A	G_B	(a) Sidesway Prevented, K	(b) Sidesway Permitted, K
AD, CF	10.0	0.346	0.781	1.754
DG, FI	0.346	0.479	0.662	1.136
BE	1.0	0.322	0.701	1.211
EH	0.322	0.446	0.654	1.130

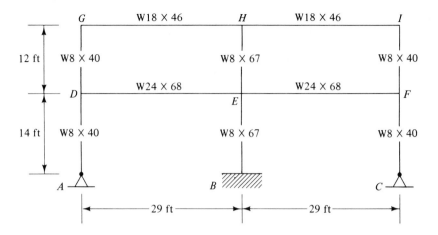

Figure 6-13   Example 6-8.

Example 6-9

A W8 × 67 section made of A36 steel is used for column *BE* of the frame shown in Fig. 6-13 of Example 6-8. It is subject to an axial load $P = 144$ k and a bending moment at the top of $M_x = 40.5$ k-ft and at the bottom of $M_x = 60.8$ k-ft, as shown in Fig. 6-14. The frame is subject to sidesway. Is the column satisfactory?

Solution.   Assume that the column is pinned and braced at the top in the weak direction. From Table 4-1(c), $K_y = 0.7$, and from Example 6-8, $K_x = 1.21$.

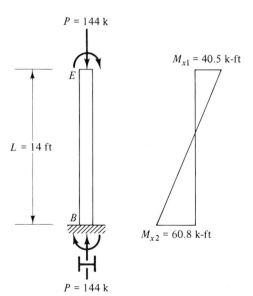

Figure 6-14   Example 6-9.

Properties of W8 × 67: $A = 19.7$ in.^{2}, $r_y = 2.12$ in., $r_x/r_y = 1.75$, $B_x = 0.326$, $a_x = 40.6(10)^6$, $L_c = 8.7$ ft, $L_u = 39.9$ ft. The effective column lengths

$$\frac{K_xL_x}{r_x/r_y} = \frac{1.21(14)}{1.75} = 9.68 \text{ ft}$$

and

$$K_yL_y = 0.7(14) = 9.8 \text{ ft}$$

For the control effective length of 9.8 ft, the W8 × 67 from the column load tables has an allowable load $P = 352$ k. The actual axial stress $f_a = P/A = 144/19.7 = 7.31$ ksi. From Table A, AISC *Manual*, p. 3-7, for $f_a = 9.6$ ksi the column is elastic; therefore, an inelastic K factor is not required. The controlling slenderness ratio

$$\frac{K_yL_y}{r_y} = \frac{0.7(14)(12)}{2.12} = 55.5$$

and the allowable axial stress is $F_a = 17.86$ ksi. Since $L_b = 14$ ft $< L_u = 39.9$ ft, $F_b = 0.6F_y = 22$ ksi. In the plane of bending

$$K_xL_x = 1.21(14)(12)$$

$$P(K_xL_x)^2 = 144(203.3)^2 = 5.95(10)^6$$

From category 1 the reduction factor $C_m = 0.85$. Substituting into Eqs. (6-7) and (6-8) [Modified AISC Formulas (1.6-1a) and (1.6-1b)], we have

$$P + P'_x = 144 + \left[0.326(60.8)(12)(0.85)\left(\frac{17.86}{22}\right) \right.$$

$$\left. \times \left(\frac{40.6(10)^6}{40.6(10)^6 - 5.95(10)^6}\right) \right]$$

$$= 336 \text{ k} < P_{all} = 352 \text{ k} \qquad \text{OK}$$

and

$$P + P'_x = 144\left(\frac{17.86}{22}\right) + \left[0.326(60.8)(12)\left(\frac{17.86)}{22}\right) \right]$$

$$= 310 \text{ k} < P_{all} = 352 \text{ k} \qquad \text{OK}$$

The W8 × 67 is satisfactory.

Example 6-10

Select a W14 section of A36 steel for the column member of the frame shown. The joints are rigid to give frame action. The frame is unbraced in the plane of bending and the columns are braced in the weak direction at the top, bottom, and midheight. Assume that there are pins in the weak direction. The loading on the columns is shown in Fig. 6-15(b).

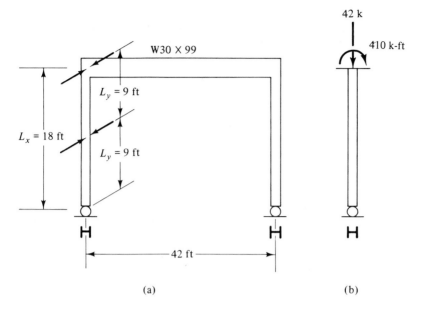

Figure 6-15   Example 6-10.

Solution.   From Eq. (6-10) the equivalent axial load

$$P_{eq} = P + 25.2 \frac{M_x}{d} = 42 + \frac{25.2(410)}{14} = 780 \text{ k}$$

Assume that $K_x = 1.8$ and $r_x/r_y = 1.6$. The effective lengths

$$\frac{K_x L_x}{r_x/r_y} = \frac{1.8(18)}{1.6} = 20.2 \text{ ft} \qquad K_y L_y = 1.0(9) = 9.0 \text{ ft}$$

From the column load tables the W14 × 159 with an effective length of 20.3 ft supports 814 k.

Calculation of K for column
W30 × 99 beam:

$$I_x = 3990 \text{ in.}^4, \ L = 42 \text{ ft}, \ I/L = 95.0 \text{ in.}^4/\text{ft}$$

W14 × 159 column:

$$I_x = 1900 \text{ in.}^4, \ L = 18 \text{ ft}, \ I/L = 105.6 \text{ in.}^4/\text{ft}$$

$$G_{top} = \frac{\sum I_c/L_c}{\sum I_b/L_b} = \frac{105.6}{95.0} = 1.11$$

$$G_{bottom} = 10.0 \qquad \text{(pinned end)}$$

The actual axial stress $f_a = P/A = 42/46.7 = 0.899 \text{ ksi} < 9.6 \text{ ksi}$. Thus from

Table A, p. 3-7 of the AISC *Manual*, the column is elastic. From the alignment chart, Fig. 6-12(b), $K = 1.93$. Properties of W14 × 159: $A = 46.7$ in.^{2}, $S_x = 254$ in.^{3}, $r_x = 6.38$ in., $r_y = 4.00$ in., $L_c = 16.4$ ft, $L_u = 57.2$ ft. The controlling slenderness ratio

$$\frac{K_x L_x}{r_x} = \frac{1.93(18)(12)}{6.38} = 65.3$$

Hence the allowable axial stress $F_a = 16.91$ ksi. The bending stress $f_b = M/S = 410(12)/254 = 19.37$ ksi, and since $L_b = 9$ ft $< L_c = 16.9$ ft, the allowable bending stress $F_b = 0.66F_y = 24$ ksi. The ratio $f_a/F_a = 0.899/16.91 = 0.053 < 0.15$. Therefore, use Eq. (6-5) [AISC Formula (1.6-2)]. Thus

$$\frac{f_a}{F_a} + \frac{f_b}{F_b} = \frac{0.899}{16.91} + \frac{19.34}{24} = 0.859 < 1.0 \qquad \text{OK}$$

The W14 × 159 is satisfactory. The W14 × 145 is also satisfactory, but the W14 × 132 is unsatisfactory. Use a W14 × 145.

Example 6-11

Select the lightest W12 section for column BC that is part of the frame shown in Fig. 6-16. The frame is unbraced in the plane of bending. The columns are braced in the weak direction at their tops and bottoms. Assume pins in the weak direction. The loading on the column and the preliminary selection of the adjacent beams is shown in the figure. Assume that the column section selected extends from A to D.

Solution. From Eq. (6-10) we make a tentative W12 selection.

$$P_{eq} = P + 25.1\frac{M_x}{d} = 500 + \frac{25.1(85)}{12} = 678 \text{ k}$$

Assume that $KL = 14$ ft. From the column load tables select a W12 × 136. The section has $r_x/r_y = 1.77$. Assume that $K_x = 2.0$, the equivalent slenderness ratio $K_x L_x/(r_x/r_y) = 2.0(14)/1.77 = 15.8$ ft. The W12 × 136 section is adequate.

Calculation of K for column BC
W24 × 55 beam:

$$I_x = 1350 \text{ in.}^4, \ L = 24 \text{ ft}, \ I/L = 56.2 \text{ in.}^4/\text{ft}$$

W12 × 136 columns:

$$I_x = 1240 \text{ in.}^4, \ L = 12 \text{ ft}, \ I/L = 103.3 \text{ in.}^4/\text{ft}$$

$$I_x = 1240 \text{ in.}^4, \ L = 14 \text{ ft}, \ I/L = 88.6 \text{ in.}^4/\text{ft}$$

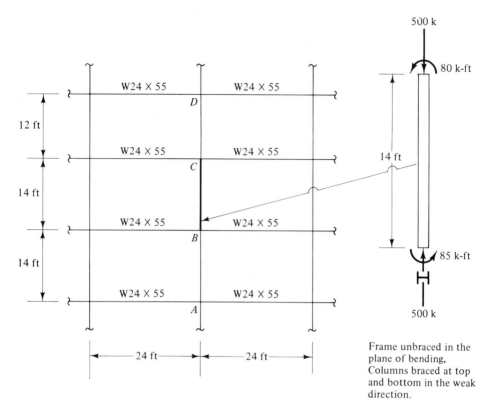

Figure 6-16   Example 6-11.

At joint B

$$G_B = \frac{\sum I_c/L_c}{\sum I_b/L_b} = \frac{2(88.6)}{2(56.2)} = 1.577$$

At joint C

$$G_C = \frac{\sum I_c/L_c}{\sum I_b/L_b} = \frac{88.6 + 103.3}{2(56.2)} = 1.707$$

The actual axial stress $f_a = P/A = 500/39.9 = 12.53$ ksi > 9.6 ksi. Therefore, a reduction in the K factor is required for inelastic effects. From Table A, p. 3-7 of the AISC *Manual*, for $f_a = 12.53$ ksi, $f_a/F'_e = 0.899$. Thus

$$G_{\text{inelastic}}(B) = \frac{f_a}{F'_e} G_{\text{elastic}}(B) = 0.899(1.577) = 1.418$$

$$G_{\text{inelastic}}(C) = \frac{f_a}{F'_e} G_{\text{elastic}}(C) = 0.899(1.707) = 1.535$$

From the alignment chart [Fig. 6-12(b)], $K = 1.45$.

Check of W12 × 136. From the column load tables:

$$L_c = 13.1 \text{ ft} < L_b = 14 \text{ ft} < L_u = 53.2 \text{ ft}$$

The section is compact; therefore, $F_b = 22$ ksi.

$$A = 39.9 \text{ in.}^2, \qquad P_{all} = 721 \text{ k}, \qquad r_y = 3.16 \text{ in.}$$

$$r_x/r_y = 1.77, \qquad B_x = 0.215, \qquad a_x = 185.1(10)^6$$

The controlling slenderness ratio

$$\frac{K_y L_y}{r_y} = \frac{1.0(14)(12)}{3.16} = 53.2$$

and the allowable axial stress $F_a = 18.07$ ksi. In the plane of bending

$$K_x L_x = 1.45(14)(12) = 243.6$$

$$P(K_x L_x)^2 = 500(243.6)^2 = 29.67(10)^6$$

The reduction factor from category 1 is $C_m = 0.85$. Substituting into Eqs. (6-7) and (6-8) [Modified AISC Formulas (1.6-1a) and (1.6-1b)], we have

$$P + P'_x = 500 + \left[0.215(85)(12)(0.85)\left(\frac{18.07}{22}\right) \right.$$

$$\left. \times \left(\frac{185.1(10)^6}{185.1(10)^6 - 29.97(10)^6} \right) \right]$$

$$= 682 \text{ k} < P_{all} = 721 \text{ k} \qquad \text{OK}$$

and

$$P + P'_x = 500\left(\frac{18.07}{22}\right) + \left[0.215(85)(12)\left(\frac{18.07}{22}\right) \right]$$

$$= 591 \text{ k} < P_{all} = 721 \text{ k} \qquad \text{OK}$$

The next lightest section, a W12 × 120, is unsatisfactory. Use a W12 × 136.

6-8 COMBINED BENDING AND AXIAL TENSION

From AISC Specification 1.6.2, members subject to combined bending and axial tension are proportioned at all points along their length to satisfy AISC Formula (1.6-1b). Thus

$$\frac{f_a}{0.6F_y} + \frac{f_{bx} \text{ (ten.)}}{F_{bx}} + \frac{f_{by} \text{ (ten.)}}{F_{by}} \le 1$$

and

$$\frac{f_b \text{ (comp.)}}{F_b} \le 1$$

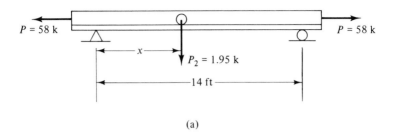

(a)

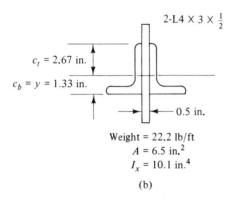

Weight = 22.2 lb/ft
$A = 6.5$ in.^{2}
$I_x = 10.1$ in.^{4}

(b)

Figure 6-17   Example 6-12.

Example 6-12

The bottom chord member of a truss shown in Fig. 6-17 is 14 ft long, has an axial tension of $P = 58$ k, and supports a movable 1.95-k differential chain hoist, including impact. The member consists of two L4 × 3 × $\frac{1}{2}$ separated by a $\frac{1}{2}$-in. gusset plate with long legs back to back. The outstanding legs serve as a track for the hoist. The connections at the end of the member are welded. Assume that the ends are pinned. Is the member satisfactory?

Solution.   The properties of two L4 × 3 × $\frac{1}{2}$ are shown in Fig. 6-17(b). The maximum moment will occur when the hoist is in the middle of the member ($x = 7$ ft). The maximum moment including the weight of the member

$$M_{\max} = \frac{PL}{4} + \frac{wL^2}{8} = \frac{1.95(14)}{4} + \frac{0.0222(14)^2}{8}$$

$$= 7.37 \text{ k-ft}$$

The actual axial tensile stress

$$f_a = \frac{P}{A} = \frac{58}{6.50} = 8.92 \text{ ksi}$$

The tensile bending stress at the bottom of the member

$$f_b \text{ (ten.)} = \frac{Mc_b}{I} = \frac{7.37(12)(1.33)}{10.1} = 11.65 \text{ ksi}$$

and the compressive bending stress at the top of the member

$$f_b \text{ (comp.)} = \frac{Mc_t}{I} = \frac{7.37(12)(2.67)}{10.1} = 23.4 \text{ ksi}$$

The allowable bending stress $F_b = 0.6F_y = 22$ ksi. From AISC Formula (1.6-1b)

$$\frac{f_a}{0.6F_y} + \frac{f_b \text{ (ten.)}}{F_b} = \frac{8.92}{22} + \frac{11.65}{22}$$

$$= 0.935 < 1.0 \qquad \text{OK}$$

and

$$\frac{f_b \text{ (comp.)}}{F_b} = \frac{23.38}{22} = 1.06 > 1.0 \qquad \text{NG}$$

The member is unsatisfactory. Two L5 × 3 × $\frac{1}{2}$ would be satisfactory.

PROBLEMS

Use the AISC *Specifications* for the following problems. In design problems determine the lightest section.

6-1. A W14 × 159 column 25 ft long of A36 steel supports an axial load $P = 500$ k and a uniformly distributed load of $w = 0.45$ k/ft as shown. The distributed load produces bending about the y axis of the column cross section. Is the column satisfactory?

6-2. A W14 × 132 column 25 ft long of A36 steel supports an axial load of $P = 500$ k and a uniformly distributed load of $w = 0.45$ k/ft as shown in Prob. 6-1. The distributed load produces bending about the x axis of the column cross section. Is the column satisfactory?

6-3. A 22-ft-long W14 × 90 beam-column of A36 steel supports an axial load $P = 220$ k and bending moments at the top of 110 k-ft and the bottom of 140 k-ft as shown. The column is braced against sidesway and pinned at the top and bottom in the x-z plane and fixed at the top and bottom in the y-z plane and sidesway is permitted. Is the column satisfactory?

6-4. A column 15 ft long supports an axial load $P = 190$ k and bending moment at the top of 80 k-ft and bottom of 90 k-ft as shown. The column is braced at midheight perpendicular to the y axis. If the column is a W10 × 77 section of A36 steel, is it satisfactory?

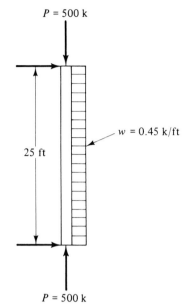

Prob. 6-1

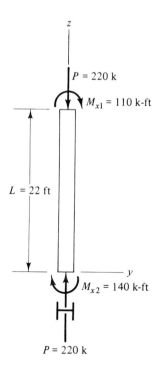

(y-z Plane: fixed top and bottom sidesway permitted)
(x-z Plane: pinned top and bottom sidesway prevented)   **Prob. 6-3**

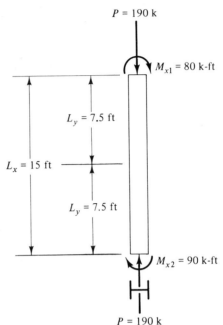

Prob. 6-4

6-5. A W14 × 74 floor beam made of A36 steel is acting as a frame member. The beam supports an axial load $P = 40$ k and a uniformly distributed load of $w = 1.7$ k/ft as shown. Lateral bracing is at the ends only. Is the beam satisfactory?

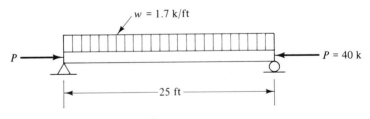

Prob. 6-5

6-6. Select the lightest W12 section section that can be used for the beam-column of Prob. 6-3.

6-7. Select the lightest W8 section of A36 steel for the member of a truss 18 ft long that supports an axial load $P = 150$ k and a transverse load $P_1 = 12$ k at midspan as shown. Lateral bracing is provided at the ends of the member and at midspan perpendicular to the y axis of the cross section. The member is part of the continuous top chord of a truss. To calculate the bending moments for the member, assume that the ends are fixed.

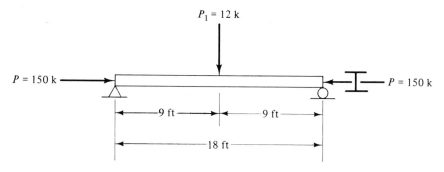

Prob. 6-7

6-8. The column shown is part of a braced frame. Select the lightest W14 section made of A36 steel to support the load $P = 300$ k with an eccentricity $e = 13$ in. with respect to the x axis of the cross section. The column height $L = 18$ ft is the effective buckling length of the member.

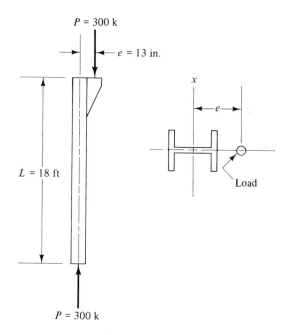

Prob. 6-8

6-9. Select the lightest W12 section of A572 Grade 50 steel that can be used in Prob. 6-5 if $P = 50$ k, $w = 1.95$ k/ft.

6-10. The column shown is part of a frame. Select the lightest W14 section made of A36 steel if the load $P = 50$ k and the moment at the top is $M = 400$ k-ft. In

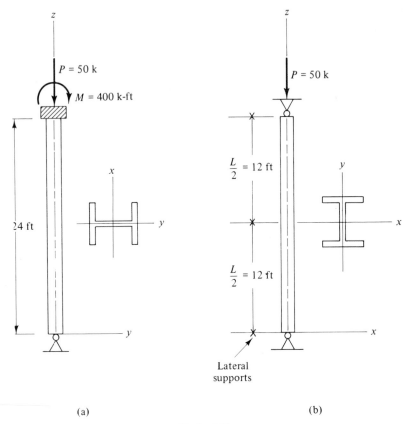

(a)   (b)

Prob. 6-10.

the y-z plane the column is fixed at the top and pinned at the bottom and sidesway
is permitted ($K_x = 2.0$). In the x-z plane the column is pinned at the top and
bottom and sidesway is prevented. Lateral supports are provided at the midheight
of the column.

6-11. Repeat Prob. 6-8 with A572 Grade 50 steel.

6-12. Repeat Prob. 6-10 with $P = 400$ k, $M = 250$ k-ft, and $L = 20$ ft.

6-13. A column of A36 steel with a length of 30 ft supports an axial load $P = 70$ k and
a moment about the x axis of the cross section $M_x = 70$ k-ft at the top end of
the column. Select the lightest W12 section if the top and bottom of the column
are fixed and no sidesway is possible.

6-14. Repeat Prob. 6-10 with $P = 90$ k, $M_x = 95$ k-ft, and A572 Grade 50 steel.

6-15. A beam-column of A36 steel with a length $L = 24$ ft supports an axial load
$P = 60$ k and a uniformly distributed load $w = 2.0$ k-ft as shown. The left end
is pinned and the right end is fixed and no sidesway is possible. Select the lightest
W14 section if
(a) Lateral supports are provided at the ends and third points.
(b) Lateral supports are provided at the ends only.

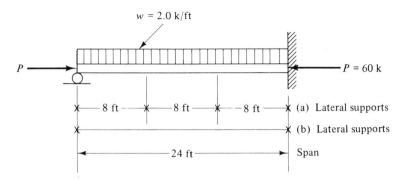

Prob. 6-15

6-16. A beam-column of A36 steel supports an axial load $P = 80$ k and a distributed load of $w = 0.090$ k/ft as shown. Both ends are pinned and no sidesway is possible. Select the lightest WT8 section if lateral supports are provided at the ends only.

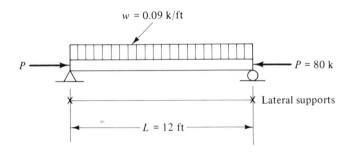

Prob. 6-16

6-17. Repeat Prob. 6-16 if the ends of the beam-column are fixed.

6-18. A column with fixed ends and length $L = 20$ ft has no sidesway or transverse loading. The compressive loads $P = 120$ are applied at the ends with eccentricities with respect to the x and y axes as shown on the cross section. Using A36 steel, determine the lightest W10 section required.

6-19. Repeat Prob. 6-18 with $P = 200$ k. Use a W14 section and A572 Grade 50 steel.

6-20. Select the lightest W12 shape for the mast of the jib crane shown in the figure. The lifting capacity including an impact allowance is 12 tons. Assume that the hoist and trolley weigh 1000 lb. Use A36 steel.

6-21. The continuous top member of the roof truss shown in the figure directly supports the roof. The roof provides continuous lateral support for the member. If a distributed load from the roof of $w = 0.75$ k/ft including the weight of the member and an axial compressive load $P_1 = 102.7$ k acts on member AB as shown, select the lightest double equal-leg angles of A36 steel that can safely support the loads. Assume that the angles are separated by $\frac{3}{8}$ in.

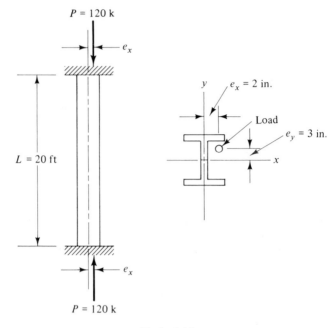

Prob. 6-18

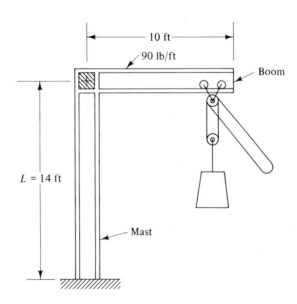

Prob. 6-20

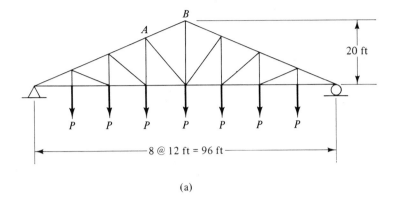

(a)

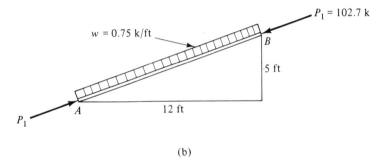

(b)

Prob. 6-21

6-22. Redesign the top member AB of Prob. 6-21 if the distributed load from the roof is $w = 1.0\,\text{k/ft}$ including the weight of the member and the axial compressive load $P_1 = 142.8\,\text{k}$.

6-23. Repeat Prob. 6-22 with a WT10.5 section.

6-24. A beam-column of A36 steel with a length of $L_x = 20\,\text{ft}$ supports an axial load $P = 470\,\text{k}$ and a moment at the top and bottom $M_x = 275\,\text{k-ft}$ as shown. The column is pinned at the top and bottom and braced against lateral buckling at midheight perpendicular to the web of the beam. Select the lightest beam-column that can safely support the loads if
  (a) A W33 section is used.
  (b) A W14 section is used.

6-25. Select a W14 section of A36 steel for the column member of the frame shown. The joints are rigid to give frame action and unbraced in the plane of bending. The columns are braced in the weak direction at the top and midheight. Assume that there are pins in the weak direction. The loading on the column is shown in the figure.

6-26. Select the lightest W12 section for column BC of the frame shown. The frame is unbraced in the plane of bending. The columns are braced in the weak direction

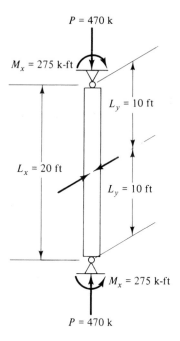

Prob. 6-24

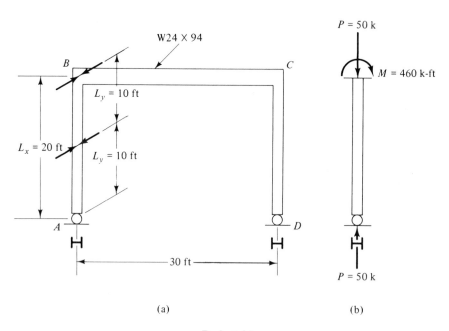

(a)   (b)

Prob. 6-25

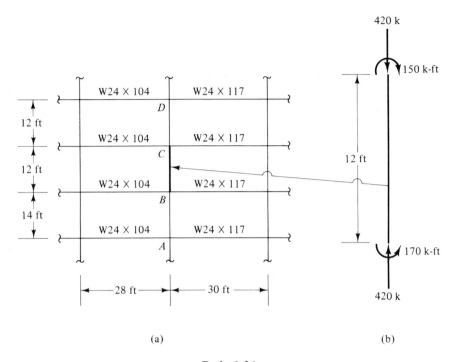

Prob. 6–26

at the top and bottom. Assume that there are pins in the weak direction. The loading on the column and the preliminary selection of the adjacent beams are shown in the figure. Assume that the column section selected extends from A to D.

6-27. Redesign the column of Prob. 6-26 if the moment at the top $M_{x1} = 75$ k-ft and at the bottom $M_{x2} = 85$ k-ft.

7

BOLTED
AND
RIVETED
CONNECTIONS

7-1 INTRODUCTION

Steel structures are made up of members formed from rolled plates and shapes that are connected together with welds or fasteners such as bolts or rivets. Welds fuse the members together, while bolts or rivets form mechanical connections between the members. For many years rivets were the principal method used for making structural connections. However, because of the economic advantages that high-strength bolts offer, they have largely replaced the rivet and unfinished bolts as connectors. In this chapter we limit ourselves mainly to the analysis and design of bolted connections. For completeness, rivets and ordinary unfinished bolts are also discussed. Welded connections are discussed in Chapter 8.

7-2 RIVETS

In recent years structural steel rivets have become virtually obsolete in the United States. Undriven rivets are manufactured so that one end has a head. The shank of the rivet is inserted while hot (between 1000 and 1950°F) into a hole that is punched or drilled $\frac{1}{16}$ in. larger than the diameter of the rivet. The other head of the rivet is formed from the shank by a pneumatic hammer as a bucking bar is pressed against the preformed head to hold the rivet in place (Fig. 7-1). During this process the rivet shank expands to almost fill the hole and press outward on the surrounding hole. As the rivet cools, both the

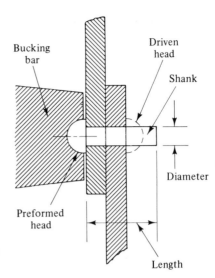

Figure 7-1 Rivet names and the riveting process.

diameter and length of the shank contract. Elastic recover of the hole tends to compensate for the decrease in the diameter of the rivet and the joined plates resist the contraction of the shank. This sets up clamping forces on the joined plates and normal force between the plates. The normal forces are unpredictable and it is customary in design calculations to neglect the resulting frictional resistance of the connection.

Rivets conform to the "Specifications for Structural Steel Rivets, ASTM A502 Grade 1 or 2." Rivet sizes range from $\frac{5}{8}$ to $1\frac{1}{2}$ in. in diameter in $\frac{1}{8}$-in. increments. Allowable stresses for rivets are given in Table 7-1 and also in the AISC *Specifications*, Table 1.5.2.1.

7-3 UNFINISHED BOLTS

Unfinished bolts are also known as ordinary or common bolts. They are used in much the same way as rivets to join shapes and plates. As in the case of rivets, the clamping force of the bolt is unpredictable and the resulting frictional resistance of the connection is neglected in design. Therefore, the design of connections with unfinished bolts is essentially the same as for rivets except for substantially lower allowable stresses. The allowable stresses are lower because of the possibility that bolt threads may extend into the shear plane.

Unfinished bolts conform to the "Specifications for Low Carbon Steel Externally and Internally Threaded Fasteners, ASTM A307." Bolt sizes are available from $\frac{1}{4}$ to 4 in. in diameter by $\frac{1}{4}$-in. increments for bolt lengths from 1 to 8 in. and by $\frac{1}{2}$-in. increments for bolt lengths over 8 in. The most commonly used bolts have hexagonal or square heads and nuts. Allowable stresses are

TABLE 7-1   Allowable Stress on Bolts and Rivets (ksi)[a]

| | | Allowable Shear, F_v | | | |
| | | Friction-type Connection[b] | | | |
Fastener	Allowable Tension, F_t	Standard Size Holes	Oversized and Short-slotted Holes	Long-slotted Holes	Bearing-type Connection
Rivets					
A502 Grade 1	23.0				17.5[c]
A502 Grade 2	29.0				22.0[c]
Bolts					
A307 (unfinished)	20.0				10.0[c]
A325-N	44.0	17.5	15.0	12.5	21.0[c]
A325-X	44.0	17.5	15.0	12.5	30.0[c]
A490-N	54.0	22.0	19.0	16.0	28.0[c]
A420-X	54.0	22.0	19.0	16.0	40.0[c]

[a]Stresses in this table to be used with a cross-sectional area are based on *nominal* diameter of connector.

[b]Stresses to be used when contact surface has *clean mill scale* (class A) conditions. (See AISC *Specifications* for Structural Joints Using ASTM A325 or A490 Bolts, Table 2a, for other surface conditions.)

[c]Reduce values 20% when length of tension splice has length between extreme fasteners, measured parallel to axial force, exceeding 50 in.

Source: Adapted from AISC *Specifications*, Table 1.5.2.1.

given in Table 7-1 and also in the AISC *Specifications*, Table 1.5.2.1. They are used chiefly on light structures under static loads or for secondary members such as bracing. AISC Specification 1.15.12 lists specific connections for which A307 bolts may *not* be used.

7-4 HIGH-STRENGTH BOLTS

These bolts, shown in Fig. 7-2(a), are made from medium-carbon steel, heat-treated, or alloy steel and have tensile strengths several times larger than those for unfinished bolts. They are tightened until they develop approximately 70 percent of the ASTM specified minimum tensile strength ($0.7F_u$). The connected plates are clamped together tightly so that most of the load transfer between plates takes place by friction, as illustrated in Fig. 7-2(b).

There are three methods for obtaining the required tension in the bolts as follows:

1. *Turn-of-the-nut method.* The bolts are first brought to a "snug," where all surfaces are in good contact and no free rotation is possible. The bolt is then given one-third to one turn, depending on the length of the bolt and slope of the surfaces on which the bolt head and nut come to rest.
2. *Calibrated wrench method.* The bolts are tightened by a manual torque wrench or power wrench that is adjusted to stall at a specified torque.
3. *Direct tension indicator.* The direct tension indicator consists of a hardened washer with protrusions on one face that flatten as the bolt is tightened. The bolt tension is measured by the remaining gap.

The turn-of-the-nut method is the most economical and reliable and is usually the preferred method.

High-strength bolts conform to the "Specifications for Structural Steel Joints Using ASTM A325 or A490 Bolts" and "Specifications for Quenched and Tempered Steel Bolts and Studs A449." ASTM A449 bolts may be used for bearing-type connections requiring bolt diameters greater than $1\frac{1}{2}$ in. or as high-strength anchor bolts and threaded rods of any diameter, AISC Specification 1.4.4. Allowable stresses for A325 and A490 bolts are given in Table 7-1 and also in AISC *Specifications*, Table 1.5.2.1.

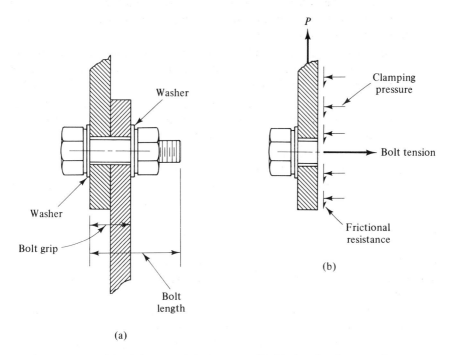

(a)

Figure 7-2   (a) High-strength bolt names; (b) frictional resistance, clamping pressure, and bolt tension.

There are three types of connections made with high-strength bolts. They are:

1. Friction-type connections (F)
2. Bearing-type connections with threads included in the shear plane (N)
3. Bearing-type connections with threads excluded from the shear plane (X)

The design of bearing type N and X connections is essentially the same as for rivets except for different allowable stresses.

7-5 TYPES OF BOLTED CONNECTIONS

To transfer loads from one structural member to another requires that connections be designed that are safe, economical, and can be fabricated. Depending on the relative position of the connected members and force being transmitted, connections are subject to various actions. The various actions are illustrated in Fig. 7-3(a) through (d) and described as follows:

(a) Axial shear occurs when the line of action of the load acts through the centroid of the bolt group. The lap joint with single shear and the butt joint with double shear are simple examples. (The small eccentricity e in the lap joint is neglected.)
(b) Eccentric shear or shear and torsion occurs when the load is applied on a line of action that does not pass through the centroid of the bolt group. A bracket attached to the flange of a column is an example of this action.
(c) Axial tension without simultaneous shear occurs when the line of action of the load is perpendicular to the member to which it is connected. A T-hanger connection is a good example.
(d) Combined axial tension and shear occur when the line of action of the load is not perpendicular to the member to which it is connected. Typical of such a connection is a diagonal brace attached to a column as shown.

7-6 METHODS OF FAILURE FOR BOLTED JOINTS

If the load on a connection is large enough to overcome friction and cause failure, the connection can fail as shown in Fig. 7-4 in any one of four ways as follows:

1. *Tension failure* of the connecting plates can occur when the plates, weakened by holes, fractures on a cross section including the holes, or when general yielding of the plates occurs. This was discussed in Chapter 3.

2. *Shear failure* of the bolt can take place in either single or double shear, depending on the number of shear planes involved. For design purposes,

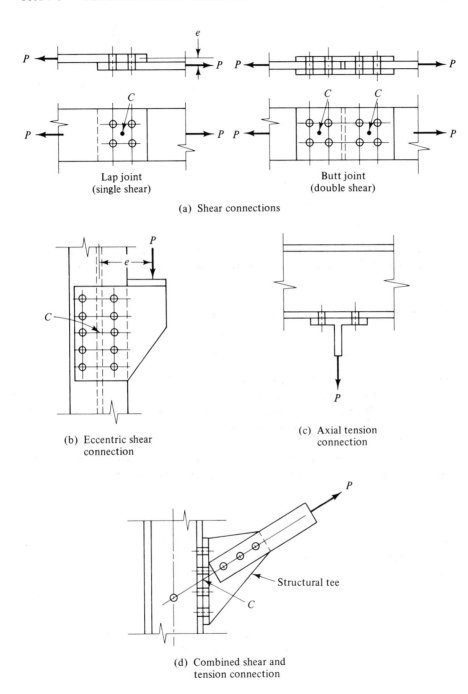

(a) Shear connections

(b) Eccentric shear connection

(c) Axial tension connection

(d) Combined shear and tension connection

Figure 7-3   Types of bolted connections.

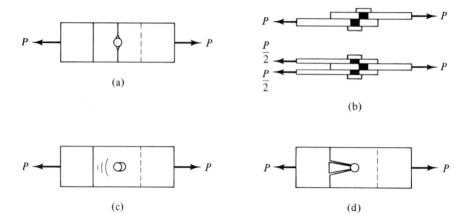

Figure 7-4   Methods of failure of bolted or riveted joints; (a) plate tension failure; (b) bolt single or double shear; (c) plate bearing failure; (d) plate tear-out.

the nominal shear stress f_v in one bolt is given by

$$f_v = \frac{P}{m(\pi D^2/4)} \tag{7-1}$$

where P = load carried by one bolt, k

D = nominal diameter of bolt, in.

m = number of shear planes that resist the load (usually one for *single shear* and two for *double shear*)

3. *Bearing failure* of the connected plates occurs when the plates crush due to the bearing force of the bolt on the material. For design purposes, the nominal bearing stress f_p can be determined from the equation

$$f_p = \frac{P}{Dt} \tag{7-2}$$

where t is the thickness of the plates (in.) and P and D are as defined for Eq. (7-1). Tests have shown that the critical or failure bearing stress $f_{p,cr}$ is given by

$$f_{p,cr} = F_u \frac{l_e}{D} \tag{7-3}$$

where F_u = tensile of ultimate strength of connecting material, ksi

l_e = distance from center of a fastener to the nearest edge of an adjacent fastener or to the free edge of a connecting part in the direction of the stress, in.

D = diameter of the fastener, in.

The center-to-edge distance l_e in terms of the center-to-center distance s is given by

$$l_e = s - \frac{D}{2}$$

Rewriting Eq. (7-3) in terms of the center-to-center distance s (spacing of the connectors), we have

$$f_{p,\text{cr}} = F_u \left(\frac{s - D/2}{D} \right) \tag{7-4}$$

Using a factor of safety of 2.0, the allowable bearing stress $F_p = f_{p,\text{cr}}/2$. From Eqs. (7-3) and (7-4), the allowable bearing stresses are given by

$$F_p = \frac{F_u l_e}{2D} \tag{7-5}$$

and

$$F_p = \frac{F_u}{2} \left(\frac{s}{D} - 0.5 \right) \tag{7-6}$$

In addition, AISC Specification 1.5.1.5.3 limits the bearing stress to a value determined from

$$F_p \leq 1.5 F_u \tag{7-7}$$

For standard holes the minimum spacing required to avoid bearing failure can be obtained by equating the actual and allowable bearing stresses from Eqs. (7-2) and (7-6) and solving for the spacing. Thus

$$s = \frac{2P}{F_u t} + \frac{D}{2} \tag{7-8}$$

$$\text{[AISC Formula (1.16-1)]}$$

Minimum spacing for standard, oversized, or slotted holes must satisfy AISC Specification 1.16.4.1 as follows:

$$s = \begin{cases} 2\frac{2}{3}D \\ \text{or} \\ 3D \quad \text{(preferred)} \end{cases}$$

4. *Tear-out* or *shear* of the connected plates can take place when the distance from the bolt to the edge of the plate in the direction of stress is too small. The minimum edge distance to avoid tear-out can be obtained by equating the actual and allowable bearing stresses from Eqs. (7-2) and (7.5) and solving for the edge distance. Thus

$$l_e = \frac{2P}{F_u t} \tag{7-9}$$

$$\text{[AISC Formula (1.16-2)]}$$

The edge distance must also satisfy AISC Specification 1.16.5.1.

Bolt spacing s and edge distance l_e may be increased to provide the required bearing stress or the bearing force may be reduced to satisfy the spacing and edge distance.

7-7 AXIALLY LOADED BOLTED AND RIVETED CONNECTIONS

In the following examples we determine the tensile capacity of various connections, the number of bolts or rivets required to develop the full tensile capacity of the member, and the minimum required value for bolt spacing and edge distance for the connection.

Example 7-1

Two $\frac{5}{8} \times 10$ in. plates made of A36 steel are joined by $\frac{7}{8}$-in.-diameter bolts or rivets as shown in Fig. 7-5. Determine the maximum allowable load on the plates if the bolts or rivets are as follows:

(a) A502 Grade 1 rivets.
(b) A325-N bolts.

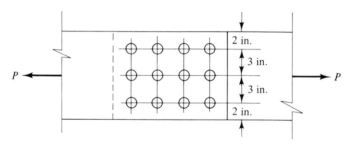

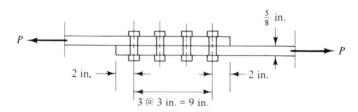

Figure 7-5   Example 7-1.

Solution.   (a) $\frac{7}{8}$-in.-diameter A502 Grade 1 rivets.

Tensile capacity. Based on the gross area:

$$F_t = 0.6F_y = 22.0 \text{ ksi} \qquad \text{(AISC Specification 1.5.1.1)}$$

$$A_g = \tfrac{5}{8}(10) = 6.25 \text{ in.}^2$$

$$P_{\text{all}} = F_t A_g = 22.0(6.25) = 137.5 \text{ k}$$

Based on the net effective area:

$$F_t = 0.5F_u = 29.0 \text{ ksi} \qquad \text{(AISC Specification 1.5.1.1)}$$

$$A_n = \left[10 - 3\left(\frac{7}{8} + \frac{1}{8}\right) \right]\left(\frac{5}{8}\right) = 4.375 \text{ in.}^2$$

$$A_e = C_t A_n = 1.0(4.375) = 4.375 \text{ in.}^2$$

$$C_t = 1.0 \qquad \text{(AISC Specification 1.14.2.2)}$$

$$A_{max} = 0.85A_g = 0.85(6.25) = 5.31 \text{ in.}^2$$

$$\text{(AISC Specification 1.14.2.3)}$$

$$P_{all} = F_t A_e = 29.0(4.375) = 126.9 \text{ k} \quad \leftarrow$$

Shear capacity. Capacity of one $\frac{7}{8}$-in. A502 Grade 1 rivet in single shear:

$$F_v = 17.5 \text{ ksi} \qquad \text{(AISC } \textit{Specification, } \text{Table 1.5.2.1)}$$

$$A_{rivet} = \frac{\pi D^2}{4} = \frac{\pi \left(\frac{7}{8}\right)^2}{4} = 0.601 \text{ in.}^2$$

$$r_v = F_v A_{rivet} = 17.5(0.601) = 10.52 \text{ k}$$

$$\text{(Also in AISC } \textit{Manual, } \text{Table I-D, p. 4-5)}$$

Capacity of 12 rivets in single shear:

$$P_{all} = r_v N = 10.52(12) = 126.2 \text{ k}$$

Bearing capacity. From Eqs. (7-5), (7-6), and (7-7), the bearing stresses for one $\frac{7}{8}$-in. rivet are given by

$$F_p = \frac{F_u l_e}{2D} = \frac{58(2)}{2(0.875)} = 66.3 \text{ ksi} \quad \leftarrow$$

$$F_p = \frac{F_u}{2}\left(\frac{s}{D} - 0.5\right) = \frac{58}{2}\left(\frac{3}{0.875} - 0.5\right) = 84.9 \text{ ksi}$$

$$F_p \leq 1.5F_u = 1.5(58) = 87 \text{ ksi}$$

Capacity of one $\frac{7}{8}$-in. rivet in bearing:

$$A_p = Dt = \left(\frac{7}{8}\right)\left(\frac{5}{8}\right) = 0.547 \text{ in.}^2$$

$$r_p = F_p A_p = 66.3(0.547) = 36.2 \text{ k}$$

Capacity of 12 rivets in bearing:

$$P_{all} = r_p N = 36.2(12) = 434.4 \text{ k}$$

The minimum allowable load is in shear. Therefore,

$$P_{max} = P_{all} \text{ (shear)} = 126.2 \text{ k}$$

(b) Tensile and bearing capacity is unchanged from part (a).

Shear capacity. Capacity of one $\frac{7}{8}$-in. A325-N bolt in single shear:

$$r_v = 12.6 \text{ k} \qquad (\text{AISC } Specifications, \text{ Table I-D, p. 4-5})$$

Capacity of 12 bolts in single shear:

$$P_{all} = r_v N = 12.6(12) = 151.2 \text{ k}$$

The minimum allowable load is in tension. Therefore,

$$P_{max} = P_{all} \text{ (tension)} = 126.9 \text{ k}$$

Example 7-2

Two $\frac{1}{2} \times 10$ in. plates made of A36 steel are joined by a butt splice as shown in Fig. 7-6. Determine the maximum allowable load on the plates if the bolts are as follows:

(a) $\frac{3}{4}$-in.-diameter A325-N bolts.
(b) $\frac{7}{8}$-in.-diameter A325-F bolts.

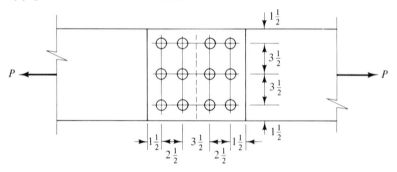

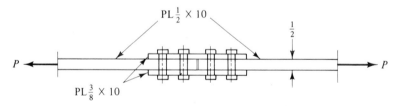

(All dimensions in inches)

Figure 7-6   Example 7-2.

Solution.   (a) The main plate will control for tension and bearing.

Tensile capacity.   Based on the gross area:

$$F_t = 0.6F_y = 22.0 \text{ ksi} \qquad (\text{AISC Specification 1.5.1.1})$$

$$P_{all} = F_t A_g = 22.0 \left(\frac{1}{2}\right)(10) = 110 \text{ k}$$

Based on the net effective area:

$$F_t = 0.5F_u = 29.0 \text{ ksi} \qquad \text{(AISC Specification 1.5.1.1)}$$

$$A_n = \left[10 - 3\left(\frac{3}{4}+\frac{1}{8}\right) \right]\left(\frac{1}{2}\right) = 3.69 \text{ in.}^2$$

$$A_e = C_t A_n = 1.0(3.69) = 3.69 \text{ in.}^2$$

$$C_t = 1.0 \qquad \text{(AISC Specification 1.14.2.2)}$$

$$A_{\max} = 0.85A_g = 0.85(5.0) = 4.25 \text{ in.}^2 \qquad \text{(AISC Specification 1.14.2.3)}$$

$$P_{\text{all}} = F_t A_e = 29.0(3.69) = 107 \text{ k} \quad \leftarrow$$

Shear capacity. Capacity of one $\frac{3}{4}$-in. A325-N bolt in double shear:

$$r_v = 18.6 \text{ k} \qquad \text{(AISC } Manual, \text{ Table I-D, p. 4-5)}$$

Capacity of six bolts in double shear:

$$P_{\text{all}} = r_v N = 18.6(6) = 111.6 \text{ k} \quad \leftarrow$$

Bearing capacity. From Eqs. (7-5), (7-6), and (7-8), the bearing stresses for one $\frac{3}{4}$-in. bolt are given by

$$F_p = \frac{F_u l_e}{2D} = \frac{58(1.5)}{2(0.75)} = 58.0 \text{ ksi} \quad \leftarrow$$

$$F_p = \frac{F_u}{2}\left(\frac{s}{D} - 0.5\right) = \frac{58}{2}\left(\frac{2.5}{0.75} - 0.5\right) = 82.2 \text{ ksi}$$

$$F_p \le 1.5F_u = 1.5(58) = 87 \text{ ksi}$$

Capacity of one $\frac{3}{4}$-in. bolt in bearing:

$$A_p = Dt = \left(\frac{3}{4}\right)\left(\frac{1}{2}\right) = 0.375 \text{ in.}^2$$

$$r_p = F_p A_p = 58.0(0.375) = 21.75 \text{ k}$$

Capacity of six bolts in bearing:

$$P_{\text{all}} = r_p N = 21.75(6) = 130.5 \text{ k} \quad \leftarrow$$

The minimum allowable load is in tension. Therefore,

$$P_{\max} = P_{\text{all}} \text{ (tension)} = 107 \text{ k}$$

(b) The main plate will control for tension and bearing.

Tensile capacity. Based on the gross area, same as for part (a):

$$P_{\text{all}} = 110 \text{ k}$$

Based on the net area:

$$F_t = 0.5F_y = 29.0 \text{ ksi}$$

$$A_e = A_n\left[10 - 3\left(\frac{7}{8} + \frac{1}{8}\right)\right]\left(\frac{1}{2}\right) = 3.5 \text{ in.}^2$$

$$C_t = 1.0 \qquad \text{(AISC Specification 1.14.2.2)}$$

$$P_{\text{all}} = F_tA_e = 29(3.5) = 101.5 \text{ k} \quad \leftarrow$$

Shear capacity. Capacity of one $\frac{7}{8}$-in. A325-F bolt in double shear:

$$r_v = 21.0 \text{ k} \qquad \text{(AISC \textit{Manual}, Table I-D, p. 4-5)}$$

Capacity of six bolts:

$$P_{\text{all}} = r_vN = 21(6) = 126 \text{ k}$$

Bearing capacity. Although the possibility of a friction-type shear connection slipping into bearing is remote, connections must still meet requirements guarding against excessive bearing stress (Commentary on AISC Specification 1.5.1.5.3). From Eqs. (7-5), (7-6), and (7-7), the bearing stresses for one $\frac{7}{8}$-in.-diameter bolt are given by

$$F_p = \frac{F_u l_e}{2D} = \frac{58(1.5)}{2(0.875)} = 49.7 \text{ ksi} \quad \leftarrow$$

$$F_p = \frac{F_u}{2}\left(\frac{s}{D} - 0.5\right) = \frac{58}{2}\left(\frac{2.5}{0.875} - 0.5\right) = 68.4 \text{ ksi}$$

$$F_p \le 1.5F_u = 1.5(58) = 87 \text{ ksi}$$

Capacity of six bolts in bearing:

$$P_{\text{all}} = F_pA_pN = 49.7\left(\frac{7}{8}\right)\left(\frac{1}{2}\right)(6) = 130.4 \text{ k}$$

The minimum allowable load is in tension. Therefore,

$$P_{\text{max}} = P_{\text{all}}\text{ (tension)} = 101.5 \text{ k}$$

Example 7-3

Determine the number of $\frac{3}{4}$-in.-diameter A325-N bolts required to join two $\frac{3}{4} \times 9$ in. plates of A36 steel with a butt splice as shown in Fig. 7-7 so that the splice can develop the full tensile capacity P_t of the member. Also determine the minimum required bolt spacing and edge distance.

Solution. The main plates will control for tension and bearing.

Tensile capacity. Based on the gross area:

$$F_t = 0.6F_y = 22.0 \text{ ksi} \qquad \text{(AISC Specification 1.5.1.1)}$$

$$P_t = F_tA_g = 22.0\left(\frac{3}{4}\right)(9) = 148.5 \text{ k}$$

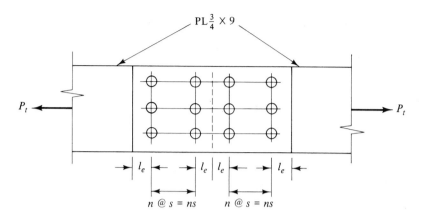

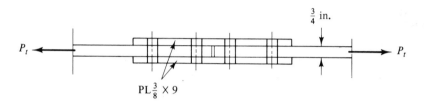

Figure 7-7   Example 7-3.

Based on the net area:

$$F_t = 0.5F_u = 29.0 \text{ ksi} \qquad \text{(AISC Specification 1.5.1.1)}$$

$$A_n = \left[9 - 3\left(\frac{3}{4} + \frac{1}{8}\right)\right]\left(\frac{3}{4}\right) = 4.78 \text{ in.}^2$$

$$C_t = 1.0 \qquad \text{(AISC Specification 1.14.2)}$$

$$A_e = A_n$$

$$A_{\max} = 0.85A_g = 0.85(6.75) = 5.74 \text{ in.}^2 \qquad \text{(AISC Specification 1.14.2.3)}$$

$$P_t = F_t A_e = 29.0(4.78) = 138.6 \text{ k} \quad \leftarrow$$

Shear capacity. The capacity of one $\frac{3}{4}$-in. A325-N bolt in double shear

$$r_v = 18.6 \text{ k} \qquad \text{(AISC } \textit{Manual,} \text{ Table I-D, p. 4-5)}$$

The number of bolts required

$$N = \frac{P_t}{r_v} = \frac{138.6}{18.6} = 7^+$$

Use nine $\frac{3}{4}$-in.-diameter A325-N bolts.

Edge distance and spacing. From Eq. (7-9), the edge distance

$$l_e = \frac{2P}{F_u t} = \frac{2(138.6/9)}{58(0.75)} = 0.708 \text{ in.}$$

From AISC Specification Table 1.16.5.1, for a $\frac{3}{4}$-in.-diameter bolt at a sheared edge $l_e = 1\frac{1}{4}$ in. Use edge distance $l_e = 1\frac{1}{4}$ in.

From Eq. (7-8), the spacing

$$s = \frac{2P}{F_u t} + \frac{D}{2} = 0.708 + \frac{0.75}{2} = 1.08 \text{ in.}$$

and from AISC Specification 1.16.4.1,

$$s = 3D = 3(0.75) = 2\frac{1}{4} \text{ in.}$$

Use spacing $s = 2\frac{1}{4}$ in.

Example 7-4

Determine the number of $\frac{3}{4}$-in. A325-X bolts required to develop the full tensile capacity P_t of the two L6 × 4 × $\frac{3}{8}$ of A36 steel as shown in Fig. 7-8. Determine the minimum required bolt spacing and edge distance for the connection.

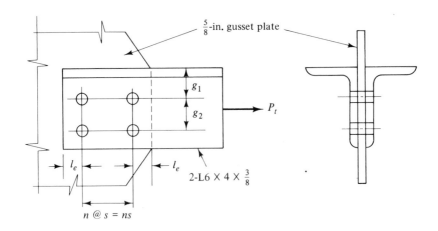

Figure 7-8   Example 7-4.

Solution

Tensile capacity. Based on the gross area:

$$F_t = 0.6F_y = 22.0 \text{ ksi} \qquad \text{(AISC Specification 1.5.1.1)}$$

$$P_t = F_t A_g = 22.0(7.22) = 158.8 \text{ k}$$

Based on the net effective area of the two angles:

$$F_t = 0.5F_u = 29.0 \text{ ksi} \qquad \text{(AISC Specification 1.5.1.1)}$$

$$A_n = A_g - n\left(d + \frac{1}{8}\right)t$$

$$= 7.22 - 4\left(\frac{3}{4} + \frac{1}{8}\right)\left(\frac{3}{8}\right) = 5.91 \text{ in.}^2$$

Assume that there are at least three fasteners per line in the direction of stress.

$$C_t = 0.85 \qquad \text{(AISC Specification 1.14.2.2)}$$

$$A_e = C_t A_n = 0.85(5.91) = 5.02 \text{ in.}^2$$

$$P_t = F_t A_e = 29.0(5.02) = 145.6 \text{ k} \quad \leftarrow$$

Shear capacity. The capacity of one $\frac{3}{4}$-in. A325-X bolt in double shear

$$r_v = 26.5 \text{ k} \qquad \text{(AISC } \textit{Manual,} \text{ Table I-D, p. 4-5)}$$

The number of bolts required

$$N = \frac{P_t}{r_v} = \frac{145.6}{26.5} = 5^+$$

Use six $\frac{3}{4}$-in.-diameter A325-X bolts.

Edge distance and spacing. Bearing on the gusset plate will control. From Eq. (7-9), the edge distance

$$l_e = \frac{2P}{F_u t} = \frac{2(145.6/6)}{58(0.625)} = 1.339 \text{ in.}$$

From AISC *Specifications,* Table 1.16.5.1 for a $\frac{3}{4}$-in.-diameter bolt at a sheared edge $l_e = 1\frac{1}{4}$ in. Use edge distance $l_e = 1\frac{1}{2}$ in.
From Eq. (7-8), the spacing

$$s = \frac{2P}{F_u t} + \frac{D}{2} = 1.339 + \frac{0.75}{2} = 1.714 \text{ in.}$$

and from AISC Specification 1.16.4.2,

$$s = 3D = 3(0.75) = 2\frac{1}{4} \text{ in.}$$

Use spacing $s = 2\frac{1}{4}$ in. The usual gage distances for an angle with a 6-in. leg from AISC *Manual,* p. 4-135, are $g_1 = 2\frac{1}{4}$ in. and $g_2 = 2\frac{1}{2}$ in.

7-8 ECCENTRICALLY LOADED BOLTED CONNECTIONS

When the line of action of the load on a connection does not pass through the centroid of the bolt group and produces only shear or bearing forces on the bolts (no tension in the shank of the bolts), the bolts are subjected to eccentric shear or shear and torsion. A typical example of eccentric shear is shown in Fig. 7-3(b), where a bracket is attached to the flange of a column.

The method of analysis for eccentrically loaded bolts discussed in the following paragraphs is based on the assumption that the behavior of the bolts is elastic. This approach has been used for many years and it provides simple and safe bolted connection. However, the designs in some cases are very conservative, with factors of safety as high as 3.0 when used with the allowable bolt stresses given in Table 7-1.

Elastic Analysis

In Fig. 7-9(a) we consider a connection with a downward force P that has an eccentricity e measured from the centroid C of the bolt group. In Fig. 7-9(b) and (c) the load is replaced by a statically equivalent system—a downward force P acting at the centroid and a torsion $M = Pe$ about the centroid.

When the load acts through the centroid each bolt supports an equal part of the load. Therefore, the force on *each* bolt [Fig. 7-10(a)] is a downward force of

$$F_y = \frac{P}{N} \tag{7-10}$$

where N is the total number of bolts in the connection. The torsion is assumed

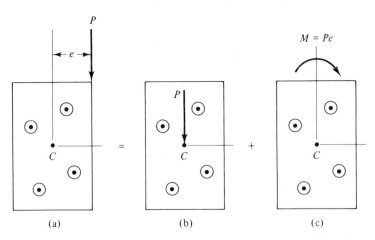

Figure 7-9 Eccentric load replaced by statically equivalent direct load and torsion.

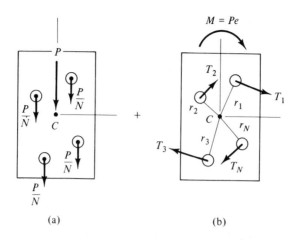

(a)   (b)

Figure 7-10   The sum of the direct force and torsion force on each bolt.

to produce forces $T_1, T_2, \ldots,$ and T_N that are normal to the distances $r_1, r_2, \ldots,$ and r_N measured from the centroid as shown in Fig. 7-10(b). Therefore,

$$M = Pe = T_1 r_1 + T_2 r_2 + \cdots + T_N r_N \qquad (7\text{-}11)$$

The forces are also assumed to be directly proportional to the distance from the centroid. Hence

$$\frac{T_1}{r_1} = \frac{T_2}{r_2} = \cdots = \frac{T_N}{r_N}$$

Solving for each of the forces in terms of T_1 and r_1 yields

$$T_1 = T_1 \frac{r_1}{r_1}, \quad T_2 = T_1 \frac{r_2}{r_1}, \quad \ldots, \quad T_N = T_1 \frac{r_N}{r_1} \qquad (a)$$

Substituting values of $T_1, T_2, \ldots,$ and T_N from Eq. (a) into Eq. (7-11) gives us

$$M = T_1 \frac{r_1^2}{r_1} + T_1 \frac{r_2^2}{r_1} + \cdots + T_1 \frac{r_N^2}{r_1}$$

or

$$M = \frac{T_1}{r_1}(r_1^2 + r_2^2 + \cdots + r_N^2) = \frac{T_1}{r_1} \sum r^2$$

Solving for the force due to torsion on bolt 1, we have

$$T_1 = \frac{M r_1}{\sum r^2} \qquad (7\text{-}12)$$

For convenience the torsion force can be resolved into x and y components as illustrated in Fig. 7-11. The x and y components of the torsion

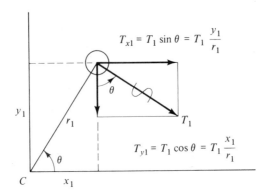

Figure 7-11   Components of the torque force.

force on bolt 1 from Eq. (7-12) and Fig. 7-11 are given by

$$T_{x1} = \frac{My_1}{\sum r^2} \qquad T_{y1} = \frac{Mx_1}{\sum r^2}$$

Similar expressions can be found for each of the bolts. For any one of the bolts

$$T_x = \frac{My}{J} \qquad T_y = \frac{Mx}{J} \qquad (7\text{-}13)$$

where $J = \sum r^2 = \sum x^2 + \sum y^2$. The directions of the components are determined by inspection. The components produce a torque about the centroid of the bolt group in the same direction as the external torque.

Example 7-5

Determine the maximum force on any one of the bolts in the bolt group shown in Fig. 7-12.

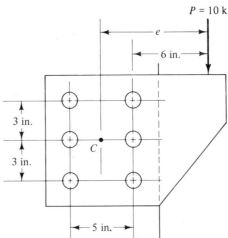

Figure 7-12   Example 7-5.

Solution.   The centroid of the bolt group is located by inspection as shown in the figure. In Fig. 7-13 we show the downward forces given by Eq. (7-10) and the horizontal and vertical torsional forces given by Eq. (7-13) on each of the bolts. The torsional forces are directed so that they produce a torque about the centroid in the same sense as the external torque.

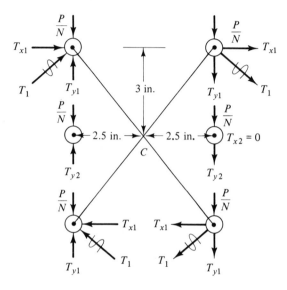

Figure 7-13   Example 7-5.

From Fig. 7-13 we see that the upper and lower bolts in the right-hand row are acted on by a maximum resultant force and in this case the forces are equal. (The maximum force is on the bolt at the end of the row nearest the line of action of the eccentric load.) From Fig. 7-12

$$e = \frac{5}{2} + 6 = 8.5 \text{ in.}$$

and the external torque

$$M = Pe = 10(8.5) = 85 \text{ k-in.}$$

The sum of the squares of the bolt group coordinates

$$\sum x^2 = 3(2.5)^2 + 3(-2.5)^2 = 37.5 \text{ in.}^2$$

and

$$\sum y^2 = 2(3)^2 + 2(-3)^2 + 2(0)^2 = 36 \text{ in.}^2$$

Therefore,

$$J = \sum x^2 + \sum y^2 = 37.5 + 36 = 73.5 \text{ in.}^2$$

For the upper bolt, from Eqs. (7-10) and (7-13),

$$F_y = \frac{P}{N} = \frac{10}{6} = 1.667 \text{ k}$$

$$T_{x1} = \frac{My_1}{J} = \frac{85(3)}{73.5} = 3.469 \text{ k}$$

and

$$T_{y1} = \frac{Mx_1}{J} = \frac{85(2.5)}{73.5} = 2.891 \text{ k}$$

The resultant of the three forces R is given by

$$R = \sqrt{T_{x1}^2 + (F_y + T_{y1})^2}$$
$$= \sqrt{(3.469)^2 + (1.667 + 2.891)^2} = 5.73 \text{ k}$$

Ultimate Strength Analysis

The ultimate strength or plastic method for eccentrically loaded bolts is based on the assumption that rotation takes place about a point called the *instant center* or instantaneous center of rotation. Deformations are assumed to be proportional to the distance from the instant center and the resulting forces are found from an exponential equation in the form

$$R = R_{\text{ult}}(1 - e^{-10\Delta})^{0.55}$$

where R = shear force on a single fastener at any given deformation
   R_{ult} = ultimate shear load on a single fastener ($F_u = 74$ ksi for an A325 bolt)
   Δ = total deformation of a fastener, including shear, bending, and bearing deformation, plus local deformation of the plate
   e = base of natural logarithm = 2.718
The exponential equation is used to fit experimentally determined load-deformation data. (The coefficients 10 and 0.55 were determined experimentally.)

Solutions by the ultimate strength method require a trial-and-error process. The location of the instant center is assumed. Based on the location of the instant center, the fastener farthest from the instant center is given a maximum deformation Δ_{max}. The maximum force for that fastener can then be calculated. Assuming a linear variation, the deformations are calculated for the remaining fasteners. With deformations known, the force for each fastener is calculated. If the assumed location of the instant center is correct, the resultant of the forces on each fastener is statically equivalent to the eccentric external load. If the assumed location of the instant center is incorrect, additional trials must be made until the correct location of the instant center has been determined.

The tables in the AISC *Manual* for eccentrically loaded connections are based on the ultimate strength analysis. Both the elastic analysis and the ultimate strength analysis (AISC *Manual*) are used in the following examples.

Example 7-6

Determine the maximum force on any one of the bolts in the bolt group shown in Fig. 7-14. Use
   (a) Elastic analysis.
   (b) Ultimate strength analysis (AISC *Manual*).

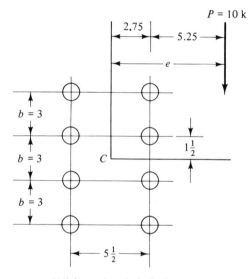

Figure 7-14   Example 7-6.   (All dimensions in inches)

Solution.   **(a)** The centroid is located by inspection as shown in the figure. A sketch of the forces on each bolt shows that the upper and lower bolts in the right-hand row are acted on by the maximum forces and in this case the forces are equal. From Fig. 7-14,

$$e = \frac{5.5}{2} + 5.25 = 8 \text{ in.}$$

Then

$$M = Pe = 10(8) = 80 \text{ k-in.}$$

$$\sum x^2 = 4(2.75)^2 + 4(-2.75)^2 = 60.5 \text{ in.}^2$$

and

$$\sum y^2 = 2(4.5)^2 + 2(1.5)^2 + 2(-4.5)^2 + 2(-1.5)^2 = 90 \text{ in.}^2$$

Therefore,

$$J = \sum x^2 + \sum y^2 = 60.5 + 90 = 150.5 \text{ in.}^2$$

For the upper bolt, from Eqs. (7-10) and (7-13),

$$F_y = \frac{P}{N} = \frac{10}{8} = 1.25 \text{ k}$$

$$T_x = \frac{My}{J} = \frac{80(4.5)}{150.5} = 2.392 \text{ k}$$

$$T_y = \frac{Mx}{J} = \frac{80(2.75)}{150.5} = 1.462 \text{ k}$$

and the resultant force

$$R = \sqrt{T_x^2 + (F_y + T_y)^2}$$
$$= \sqrt{(2.392)^2 + (1.25 + 1.462)^2} = 3.61 \text{ k}$$

(b) From the table for eccentric loads on fastener groups, Table XII, AISC *Manual*, p. 4-64, for $b = 3$ in, $l = e = 8$ in., and $n = 4$: $C = 3.30$. Therefore, the maximum force on a connector with $P = 10$ k is given by

$$R = r_v = \frac{P}{C} = \frac{10}{3.30} = 3.03 \text{ k}$$

In this example the elastic analysis gives a result that is approximately 20 percent larger than the ultimate strength analysis. The usual percentages are 20 percent or less. The AISC *Specifications* continue to recognize the elastic method for fastener groups that are not tabulated in the AISC *Manual*.

Example 7-7

Determine the number of bolts required to connect the eccentrically loaded bracket to the column flange as shown in Fig. 7-15. Use A36 steel and $\frac{7}{8}$-in.-diameter
  (a) A325-F bolts.
  (b) A325-N bolts.

Solution. **(a)** $\frac{7}{8}$-in. A325-F bolts.

Shear capacity. Capacity of one bolt in single shear:

$$r_v = 10.5 \text{ k} \qquad \text{(AISC *Manual*, Table I-D, p. 4-5)}$$

Bearing capacity. From Eqs. (7-5), (7-6), and (7-7), the bearing stresses

$$F_p = \frac{F_u l_e}{2D} = \frac{58(1.5)}{2(0.875)} = 49.7 \text{ ksi} \quad \leftarrow$$

$$F_p = \frac{F_u}{2}\left(\frac{s}{D} - 0.5\right) = \frac{58}{2}\left(\frac{3}{0.875} - 0.5\right) = 84.9 \text{ ksi}$$

$$F_p \le 1.5 F_u = 1.5(58) = 87 \text{ ksi}$$

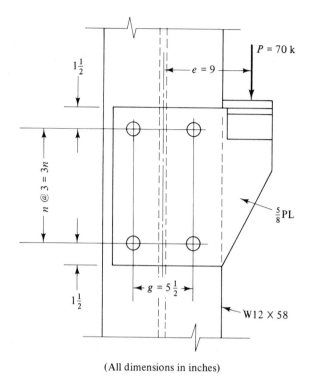

(All dimensions in inches)

Figure 7-15   Example 7-7.

The flange thickness $t = 0.640$ in.; therefore, the bearing capacity will be based on the $\frac{5}{8}$-in. plate.

$$A_p = Dt = \left(\frac{7}{8}\right)\left(\frac{5}{8}\right) = 0.547 \text{ in.}^2$$

$$r_p = F_p A_p = 49.7(0.547) = 27.2 \text{ k}$$

Shear controls. For $P = 70$ k and $r_v = 10.5$ k,

$$C = \frac{P}{r_v} = \frac{70}{10.5} = 6.67$$

From the table for eccentric loads on fastener groups, Table XII, AISC *Manual*, p. 4-64, for $b = 3$ in., $l = e = 9$ in., and $C = 6.67$: $n = 7$ bolts in each vertical row. Use two rows of bolts with seven $\frac{7}{8}$-in. A325-X bolts in each row.

Check. The bending and shear stress in the bracket plate are checked as follows. The depth of the bracket plate is

$$2(1.5) + 6(3) = 21 \text{ in.}$$

The stresses in the bracket are complex and simple bending theory does not apply. However, it will furnish a rough check. The net section modulus through the gage line nearest the load from the AISC *Manual*, p. 4-87, is $S = 31$ in.^{3}. The maximum bending stress on the section through the gage line is given by

$$f_b = \frac{M}{S} = \frac{70[9 - (5.5/2)]}{31} = 14.1 \text{ ksi} < 0.6F_y = 22 \text{ ksi} \qquad \text{OK}$$

The shear area of the net section

$$A_v = \left[21 - 7\left(\frac{7}{8} + \frac{1}{8}\right)\right]\left(\frac{5}{8}\right) = 8.75 \text{ in.}^2$$

and the shear stress

$$f_v = \frac{P}{A_v} = \frac{70}{8.75} = 8.0 \text{ ksi} < 0.4F_y = 14.4 \text{ ksi} \qquad \text{OK}$$

(b) $\frac{7}{8}$-in. A325-N bolt.

Shear capacity. The capacity of one bolt in single shear

$$r_v = 12.6 \text{ k} \qquad \text{(AISC } \textit{Manual, } \text{Table I-D, p. 4-5)}$$

Bearing capacity. Same as for part (a); therefore, shear controls. For $P = 70$ k and $r_v = 12.6$ k,

$$C = \frac{P}{r_v} = \frac{10}{12.6} = 5.56$$

From the table for eccentric loads on fastener groups, Table XII, AISC *Manual*, p. 4-64, for $b = 3$ in., $l = e = 9$ in., and $C = 5.56$: $n = 6$ bolts in a vertical row. Use two rows of bolts with six $\frac{7}{8}$-in. A325-N bolts in each row.

The bending and shear stresses in the bracket plate were checked and found to be satisfactory.

7-9 CONNECTIONS SUBJECT TO SHEAR AND TENSION

Tests have shown that the *strength* of bearing-type fasteners subject to both shear and tension can be approximated by an equation in the form of an ellipse. After applying a factor of safety, the equation can be written as

$$\left(\frac{f_v}{F_v}\right)^2 + \left(\frac{f_t}{F_t}\right)^2 = 1$$

where f_v = average shear stress in all the bolts
$\quad f_t$ = average tensile stress in all the bolts
$\quad F_v$ = allowable shear stress if shear forces alone exist
$\quad F_t$ = allowable tensile stress if tensile forces alone exist

For bearing-type connections, AISC Specification 1.6.3 uses three straight lines to approximate the ellipse. In friction-type connections the AISC *Specifications* use a more conservative single, straight-line relationship. The equations are given in Table 7-2.

TABLE 7-2   Allowable Combined Tension and Shear Stresses for Bolted Connections[a]

Fastener	Allowable stresses: F'_t, F_v (ksi)
A307 bolt	$F'_t = 26 - 1.8f_v \leq 20$; $f_v = F_v = 10$
A325-N bolt	$F'_t = 55 - 1.8f_v \leq 44$; $f_v = F_v = 21$
A325-X bolt	$F'_t = 55 - 1.4f_v \leq 44$; $f_v = F_v = 30$
A490-N bolt	$F'_t = 68 - 1.8f_v \leq 54$; $f_v = F_v = 28$
A490-X bolt	$F'_t = 68 - 1.4f_v \leq 54$; $f_v = F_v = 40$
A325-F bolt	$F'_v = 17.5\left(1 - \dfrac{f_t A_b}{T_b}\right)$ bolt in standard hole[b]
A490-F bolt	$F'_v = 22.0\left(1 - \dfrac{f_t A_b}{T_b}\right)$ bolt in standard hole[b]

[a] F'_t, allowable tensile stress when combined with shear stress in a bearing-type connection; F'_v, allowable shear stress when combined with tensile stress in a shear-type connection; f_v, average shear stress in all the bolts; f_t, average tensile stress in all the bolts; A_b, nominal area of the fastener; T_b, specified pretension load in the bolt (listed in AISC *Manual*, Table 1.23.5).
[b] For oversize or slotted holes, replace the values of the allowable shear stress $F_v = 17.5$ ksi and $F_v = 22.0$ ksi by values from AISC Specification 1.5.2.2.
Source: Adapted from AISC Specifications 1.6.3 and 1.5.2.2. (Courtesy of the American Institute of Steel Construction, Inc.)

Example 7-8

Determine if the connection shown in Fig. 7-16 is adequate when eight
   (a) $\frac{3}{4}$-in. A325-N bolts are used.
   (b) $\frac{7}{8}$-in. A325-F bolts are used.
Disregard the effect of prying.

Solution

$$\text{Tension component: } T = 0.8(100) = 80 \text{ k}$$
$$\text{Shear component: } \quad V = 0.6(100) = 60 \text{ k}$$

(a) For eight $\frac{3}{4}$-in. A325-N bolts. The allowable shear stress

$$F_v = 21.0 \text{ ksi} \qquad \text{(Table 7-2)}$$

and the bolt shear stress

$$f_v = \frac{V}{NA_b} = \frac{60}{8(0.442)} = 17.0 \text{ ksi} < F_v = 21.0 \text{ ksi} \qquad \text{OK}$$

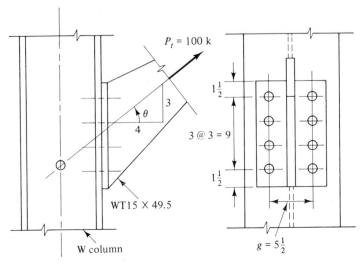

(All dimensions in inches)

Figure 7-16   Example 7-8.

The allowable tensile stress from Table 7-2

$$F_t' = 55 - 1.8 F_v = 55 - 1.8(17.0) = 24.4 \, \text{ksi}$$

and the tensile stress in the bolts

$$f_t = \frac{T}{NA_b} = \frac{80}{8(0.442)} = 22.6 \, \text{ksi} < F_t' = 24.4 \, \text{ksi} \qquad \text{OK}$$

The connection is adequate.

    (b) For eight $\frac{7}{8}$-in. A325-F bolts. The allowable tensile stress

$$F_t = 44.0 \, \text{ksi} \qquad (\text{AISC } \textit{Manual}, \text{Table I-A})$$

and the bolt tensile stress

$$f_t = \frac{T}{NA_b} = \frac{80}{8(0.601)} = 16.6 \, \text{ksi} < F_t = 44.0 \, \text{ksi} \qquad \text{OK}$$

The minimum bolt tension $T_b = 39$ k (AISC *Specifications*, Table 1.23.5). From Table 7-2 the allowable shear stress

$$F_v' = 17.5 \left(1 - \frac{f_t A_b}{T_b} \right) = 17.5 \left(1 - \frac{16.6(0.601)}{39} \right) = 13.0 \, \text{ksi}$$

and the bolt shear stress

$$f_v = \frac{V}{NA_b} = \frac{60}{8(0.601)} = 12.5 \, \text{ksi} < F_v' = 13.0 \, \text{ksi} \qquad \text{OK}$$

The connection is adequate. The prying forces on the bolts should also be investigated.

Example 7-9

Determine the number of bolts required if the tensile load $P_t = 115$ k and the slope $\theta = 45°$ for the connection shown in Fig. 7-16. Use $\frac{7}{8}$-in.-diameter
  (a) A490-N bolts.
  (b) A490-F bolts.
Disregard the effect of prying.

Solution.   The tensile and shear components

$$T = V = 115 \cos 45° = 81.3 \text{ k}$$

(a) Bearing-type connection. The interaction equation for an A490-N bolt from Table 7-2 is

$$F_t' = 68 - 1.8 f_v \le 54 \text{ ksi} \qquad \text{(a)}$$

and the allowable shear stress

$$F_v = 28 \text{ ksi}$$

Changing Eq. (a) to a force equation by multiplying by the area of the bolts NA_b, we have

$$F_t' NA_b = 68 NA_b - 1.8 f_v NA_b \le 54 NA_b$$

Then since

$$T = 81.3 \text{ k} = F_t' NA_b \qquad \text{and} \qquad V = 81.3 \text{ k} = f_v NA_b$$

we have

$$81.3 = 68 NA_b - 1.8(81.3) \le 54 NA_b \qquad \text{(b)}$$

Solving for NA_b from Eq. (b) yields

$$NA_b = \frac{81.3 + 1.8(81.3)}{68} = 3.35 \text{ in.}^2$$

This satisfies the tension requirement. The shear must be checked.

$$f_v = \frac{V}{NA_b} = \frac{81.3}{3.35} = 24.3 \text{ ksi} < F_v = 28 \text{ ksi} \qquad \text{OK}$$

For $\frac{7}{8}$-in.-diameter bolts, $A_b = 0.601$ in.^{2}. Thus the number of bolts required

$$N = \frac{NA_b}{A_b} = \frac{3.35}{0.601} = 5^+$$

Use six $\frac{7}{8}$-in.-diameter A490-N bolts.
    The prying forces should also be investigated.

(b) Friction-type connection. The interaction equation for an A490-F bolt from Table 7-2 is

$$F_v' = 22.0\left(1 - \frac{f_t A_b}{T_b}\right) \tag{c}$$

The minimum bolt tension for a $\frac{7}{8}$-in. A490-F bolt is $T_b = 49$ k (AISC *Specifications*, Table 1.23.5). Then $A_b/T_b = 0.601/49 = 0.01227$. Substituting into Eq. (c), we have

$$F_v' = 22.0 - 22.0(0.01227)f_t$$
$$= 22.0 - 0.270f_t \tag{d}$$

Changing Eq. (d) to a force equation by multiplying by the area of the bolts NA_b, we have

$$F_v' NA_b = 22NA_b - 0.270f_t NA_b$$

Then since

$$V = 81.3 \text{ k} = F_v' NA_b \quad \text{and} \quad T = 81.3 \text{ k} = f_t NA_b$$

we have

$$81.3 = 22NA_b - 0.270(81.3) \tag{e}$$

solving for NA_b from Eq. (e) gives us

$$NA_b = \frac{81.3 + 0.270(81.3)}{22} = 4.61 \text{ in.}^2$$

This satisfies the shear requirement. The tension stress must be checked.

$$f_t = \frac{T}{NA_b} = \frac{81.3}{4.61} = 17.3 \text{ ksi} < F_t = 54 \text{ ksi} \quad \text{OK}$$

For $\frac{7}{8}$-in.-diameter bolts, $A_b = 0.601$ in.^{2}. Thus the number of bolts required

$$N = \frac{NA_b}{A_b} = \frac{4.61}{0.601} = 7^+$$

Use eight $\frac{7}{8}$-in.-diameter A490-F bolts.

The prying forces should also be investigated.

7-10 BUILDING FRAME CONNECTIONS

Connections are classified in AISC Specification 1.2 according to their rotational characteristics under load as rigid frame (Type 1), simple framing (Type 2), and semirigid framing (Type 3). Connections that develop the full moment capacity and maintain a constant angle between the connected members are called rigid connections, while connections that develop no resistive moment and are free to rotate are called simple connections. The semirigid connection develops less than the full moment capacity of the connected members.

Actual connections are neither completely rigid nor completely flexible and can be classified on the basis of the *ratio of the moment developed by the connection to the full moment capacity of the connected member expressed as a percentage.* The percentage for a particular connection must be determined by actual test. The approximate percentages for a simple connection are from 0 to 20, for a semirigid connection from 20 to 80, and for a rigid connection from 80 to 90. Typical moment-rotation curves for the three types of connections on a uniformly loaded beam are displayed in Fig. 7-17.

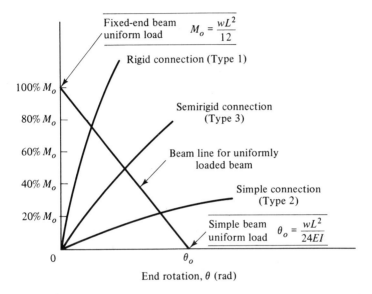

Figure 7-17 Typical end moment-rotation curves for AISC connection types (uniformly loaded beam).

7-11 SHEAR CONNECTIONS FOR BUILDING FRAMES

The shear or web-framing connection is used to connect beams to girders or columns and provide simple supports. They have been the most commonly used method of framing for small buildings where the connections are simple connections (AISC Type 2). In the shear connection the web of the beam is connected by two framing web angles to the girder or column as shown in Fig. 7-18. The framing angles are made thin so they are as flexible as possible. Most of the flexibility of the connection is due to the bending and twisting of the angles. However, the connections do develop moments that may be as large as 20 percent of the full fixed-end moment. The moments are not considered in design.

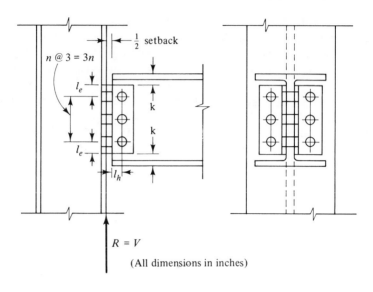

Figure 7-18   Double-angle bolted web-framing connection.

The design procedure for shear connections is illustrated in the following examples.

Example 7-10

Determine the number of bolts required to connect a W27 × 94 beam to a W12 × 152 column with a double-angle web-framing connection and the minimum horizontal edge distance l_h for the beam (Fig. 7-18). Rolled sections are of A36 steel. The reactive force to be transmitted is $R = 90$ k. Use $\frac{3}{4}$-in.-diameter

   (a) A325-F bolts in standard holes.
   (b) A325-N bolts in standard holes.

Solution.   **(a)** Use $\frac{3}{4}$-in.-diameter A325-F bolts. The W27 × 94 beam has a web thickness $t_w = 0.490$ in. The suggested thickness of the web angle with $\frac{3}{4}$-in.-diameter A325-F bolts from AISC *Manual*, Table II-A, p. 4-24, is $\frac{1}{4}$ in.

   Angle connection to web of beam. The shear capacity of the bolts in double shear $r_v = 15.5$ k (AISC *Manual*, Table I-D) and the bearing capacity of the two $\frac{1}{4}$-in.-thick angles with the usual 3-in. spacing $r_p = 32.6$ k (AISC *Manual*, Table I-E). The number of bolts required based on shear

$$N = \frac{R}{r_v} = \frac{90}{15.5} = 5^+ \quad \text{(use 6 bolts)}$$

   Angle connection to the flange of the column. The shear will control. The capacity of the bolts in single shear $r_v = 7.7$ k (AISC *Manual*, Table I-D). The number of bolts required

$$N = \frac{R}{r_v} = \frac{90}{7.7} = 11^+ \quad \text{(use 12 bolts)}$$

For an edge distance $l_e = 1\frac{1}{4}$ in. (AISC *Manual*, Table 1.16.5.1) the total length of the web angle with 3-in. bolt spacing

$$l = 2(1.25) + 5(3) = 17.5 \text{ in.}$$

For the web angles the shear area

$$A_v = 2\left[17.5 - 6\left(\frac{3}{4} + \frac{1}{8}\right) \right]\left(\frac{1}{4}\right) = 6.313 \text{ in.}^2$$

and the shear stress

$$f_v = \frac{R}{A_v} = \frac{90}{6.313}$$

$$= 14.3 \text{ ksi} < 0.3 F_u = 0.3(58) = 17.4 \text{ ksi} \qquad \text{OK}$$
$$\text{(AISC Specification 1.5.1.2.2)}$$

The minimum horizontal edge distance

$$l_h = \frac{2P_r}{F_u t} = \frac{2(90/6)}{58(0.490)} = 1.06 \text{ in.} \qquad \text{(AISC Specification 1.16.5.3)}$$

Use $l_h = 1\frac{1}{4}$ in.

Solution by Tables (AISC *Manual*). For W27 $\times$ 94, $t_w = 0.490$ in. and reaction $R = 90$ k.

Angle thickness and number of bolts required. Table II-A: For $\frac{3}{4}$-in. A325-F bolts the suggested angle thickness $t = \frac{1}{4}$ in. and with $N = 6$ bolts the angle has a length $l = 17.5$ in. and a shear capacity of 92.8 k > 90 k. (OK)

Shear capacity of angles. Table II-C: The allowable shear capacity for a $\frac{1}{4}$-in. angle if six $\frac{3}{4}$-in. bolts are used is 110 k > 90 k. (OK) (This check is not required. Table II-A indicates when net shear on the angle is critical.)

Bearing capacity of web. Table I-E: For $F_u = 58$ ksi, $\frac{3}{4}$-in. bolts, and the usual 3-in. spacing, the allowable load for 1-in.-thick material is 65.3 k/bolt. For $t_w = 0.490$ in. and six bolts, the allowable bearing capacity of the web is $6(0.490)(65.3) = 192.0$ k > 90 k. (OK)

Bearing capacity of two $\frac{1}{4}$-in. angles. Table I-E: For $F_u = 58$ ksi, $\frac{3}{4}$-in. bolts, and the usual 3-in. spacing, the allowable load for $\frac{1}{4}$-in.-thick material is 16.3 k/bolt. For six bolts in each angle the capacity is $6(2)(16.3) = 195.6$ k > 90 k. (OK)

Edge distance. Table I-F: For $l_e = l_h = 1\frac{1}{4}$ in. and $F_u = 58$ ksi, the allowable load for 1-in.-thick material is 36.3 k/bolt. For six bolts and a web thickness $t_w = 0.490$ in., the capacity is $6(0.490)(36.3) = 106.7$ k > 90 k. (OK)

Use six $\frac{3}{4}$-in.-diameter A325-F bolts, $\frac{1}{4}$-in.-thick web angles, 3-in. bolt spacing, and $l_e = l_h = 1\frac{1}{4}$ in.

(b) Use $\frac{3}{4}$-in.-diameter A325-N bolts.

Solution by Tables (AISC *Manual*).   For W27 × 94, $t_w = 0.490$ in. and reaction $R = 90$ k.

Angle thickness and number of bolts required. Table II-A: For specified bolts the suggested angle thickness $t = \frac{5}{16}$ in. with $N = 5$ bolts has a length $l = 14.5$ in. and a shear capacity of 92.8 k > 90 k.   (OK)

Shear capacity of angles. No check is required (see Table II-A).

Bearing capacity of web. Same as part (a): 65.3 k/bolt/1-in. thickness. For $t_w = 0.490$ in. and five bolts, the allowable bearing capacity of the web is $5(0.490)(65.3) = 160$ k > 90 k.   (OK)

Bearing capacity of two $\frac{5}{16}$-in. angles. Table I-E: For $F_u = 58$ ksi, $\frac{3}{4}$-in.-diameter bolts, and the usual 3-in. spacing, the allowable load for $\frac{5}{16}$-in.-thick material is 20.4 k/bolt. For five bolts the capacity is $5(2)(20.4) = 204$ k > 90 k.   (OK)

Edge distance. Same as part (a), $l_e = 1\frac{1}{4}$ in. Increase horizontal edge distance l_h to $1\frac{1}{2}$ in. Table I-F: For $l_h = 1\frac{1}{2}$ in. and $F_u = 58$ ksi, the allowable load for 1-in.-thick material is 43.5 k/bolt. The capacity for five bolts on a web thickness $t_w = 0.490$ in. is $5(0.490)(43.5) = 106.6$ k > 90 k.   (OK)

Use five A325-N bolts, $\frac{5}{16}$-in. web angles, 3-in. spacing, and edge distances $l_e = 1\frac{1}{4}$ in. and $l_h = 1\frac{1}{2}$ in.

Example 7-11

Determine the number of bolts required to connect a W18 × 55 beam to a W33 × 141 girder with a double-angle web-framing connection as shown in Fig. 7-19. Rolled sections are A36 steel. The reactive force to be transmitted is $R = 85$ k. The flange of the beam will be coped. Use $\frac{3}{4}$-in.-diameter A325-N bolts in standard holes.

Solution by Tables (AISC *Manual*).   For a W18 × 55 beam, $t_w = 0.390$ in., a reaction $R = 85$ k, and $\frac{3}{4}$-in.-diameter A325-N bolts.

Angle thickness and number of bolts required. Table II-A: For the specified bolts the suggested angle thickness $t = \frac{5}{16}$ in. with $N = 5$ bolts has a length of $l = 14.5$ in. and a shear capacity of 92.8 k > 85 k.   (OK)

Shear capacity of angles. No check is required (see Table II-A).

Bearing capacity of web. Table I-E: For $F_u = 58$ ksi, $\frac{3}{4}$-in.-diameter bolts, and the usual 3-in. spacing, the allowable load for 1-in. material is 65.3 k/bolt. For $t_w = 0.390$ in. and five bolts, the allowable bearing capacity of the web is $5(0.390)(65.3) = 127.3$ k > 85 k.   (OK)

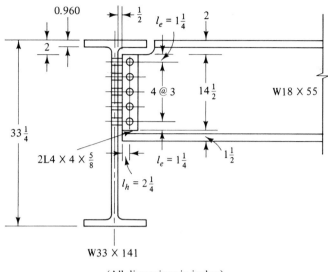

W33 × 141

(All dimensions in inches)

Figure 7-19   Example 7-11.

Web tear-out (block shear). Assume that $l_v = 1\frac{1}{4}$ in. and $4 \times 4 \times \frac{5}{16}$ angles are used. Then

$$l_h = 4 - \tfrac{1}{2} - 1\tfrac{1}{4} = 2\tfrac{1}{4} \text{ in.}$$

Table I-G: For $l_v = 1\frac{1}{4}$ in. and $l_h = 2\frac{1}{4}$ in., $C_1 = 1.50$, and for $N = 5$ and bolt diameter of $\frac{3}{4}$ in., $C_2 = 2.30$. Therefore, the beam tear-out resistance

$$(C_1 + C_2)F_u t = (1.50 + 2.30)(58)(0.390)$$
$$= 85.9 \text{ k} > 85 \text{ k} \qquad \text{OK}$$

Number of bolts required to connect angles to girder. Table I-D: Shear capacity of $\frac{3}{4}$-in. A325-N bolts in single shear, $r_v = 9.3$ k. The number of bolts required

$$N = \frac{R}{r_v} = \frac{85}{9.3} = 9^+ \quad \text{(use 10 bolts)}$$

Bearing on the angles with a thickness of $t = \frac{5}{16}$ in. would control rather than the girder web with a thickness $t_w = 0.605$ in.

Table I-E: For $F_u = 58$ ksi, $\frac{3}{4}$-in.-diameter bolts, and the usual 3-in. spacing, the allowable load for $\frac{5}{16}$-in.-thick material is 20.4 k/bolt. For 10 bolts the allowable bearing capacity is 10(20.4) = 204 k > 85 k.   (OK)

See Fig. 7-19 for details of connection.

7-12 SEATED BEAM CONNECTIONS

The seated beam connection shown in Fig. 7-20 can be used to provide simple support for the ends of beams. It consists of an unstiffened seat (an angle) that is designed to support the total reaction and a top clip angle that provides lateral support for the compression flange. The angles are made as thin and flexible as practical and under load the outstanding legs of both angles are bent downward. Although designed to transmit only a vertical reaction the connections do develop moments that may be as large as 30 percent of the full fixed-end moment. This moment is not considered in design.

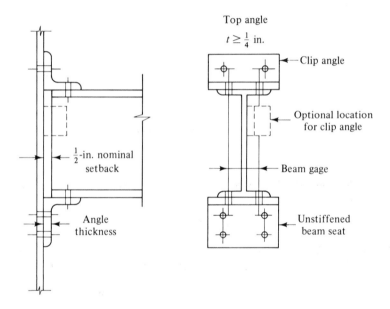

Figure 7-20   Unstiffened seated beam connection.

The thickness of the angle is determined by bending stress on a critical section of the angle as shown in Fig. 7-21. Since the distribution of forces, the location of the resultant reaction, and the critical section for bending of the seat angle are all unknown, the locations shown in the figure are design assumptions. The assumptions here are those used for the tables in the AISC *Manual.* The various design assumptions are discussed in the following example.

Example 7-12

Design a seated beam connection to support a W18 × 40 beam of A36 steel. The reactive force is $R = 26$ k. The beam seat is connected to a column web with a gage of $5\frac{1}{2}$-in. by $\frac{7}{8}$-in.-diameter A325-F bolts.

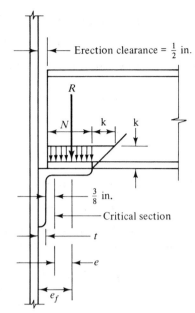

Erection clearance $= \frac{1}{2}$ in.

R

N k k

$\frac{3}{8}$ in.

Critical section

t

e

e_f

Figure 7-21 Terms for an unstiffened beam seat.

Solution. W18 × 40: $t_w = 0.315$ in., $k = 1\frac{3}{16} = 1.375$ in. The required bearing length N is determined from the stress in the web of the beam and is found from Eq. (5-9) [AISC Formula 1.10-9]. Solving for N, we have

$$N = \frac{R}{0.75 F_y t} - k = \frac{26}{0.75(36)(0.315)} - 1.375$$

$$= 1.682 \text{ in.}$$

The nominal erection clearance is $\frac{1}{2}$ in. To provide for a possible mill underrun in beam length, we assume a $\frac{3}{4}$-in. clearance. From Fig. 7-21,

$$e_f = \text{erection clearance} + \frac{N}{2}$$

$$= \frac{3}{4} + \frac{1.682}{2} = 1.591 \text{ in.} \tag{a}$$

Assume that the angle thickness $t = \frac{3}{4}$ in. Then

$$e = e_f - t - \frac{3}{8}$$

$$= 1.591 - 0.75 - 0.375 = 0.466 \text{ in.} \tag{b}$$

The critical cross section is rectangular with a width b and a thickness t. The allowable bending stress for a solid rectangular section bent about its weaker axis is given by AISC Specification 1.5.1.4.3 as

$$F_b = 0.75F_y$$

Therefore, from the bending stress formula

$$f_b = \frac{Mc}{I} = \frac{Mt/2}{bt^3/12} \leq F_b = 0.75F_y$$

Solving for t^2, we have

$$t^2 \geq \frac{6M}{0.75F_y b} \qquad\qquad (c)$$

The bending moment on the critical section is given by

$$M = Re = 26(0.466) = 12.12 \text{ k-in.}$$

With a gage $g = 3\frac{1}{2}$ in. the usual length of seat $b = 6$ in. With a gage $g = 5\frac{1}{2}$ in. the usual length $b = 8$ in. Use 8 in. The angle thickness from Eq. (c)

$$t^2 = \frac{6(12.12)}{0.75(36)(8)} = 0.337$$

$$t = 0.580 \text{ in.} < 0.75 \text{ in.} \qquad \text{OK}$$

The A325-F bolt with a diameter of $\frac{7}{8}$ in. has a single shear capacity $r_v = 10.5$ k (AISC *Manual*, Table I-D). The number of bolts required $N = R/r_v = 26/10.5 = 2^+$. Use three $\frac{7}{8}$-in. A325-F bolts and seat angle L4 × 4 × $\frac{3}{4}$ × 0 ft-8 in.

Example 7-13

Design the seated connection to support a W21 × 50 beam of A36 steel. The reactive force is $R = 30$ k and the beam seat is connected to a column flange with a gage of $5\frac{1}{2}$-in. by $\frac{7}{8}$-in.-diameter A325-N bolts.

Solution by tables (AISC *Manual*). With a gage of $5\frac{1}{2}$ in. we use a seat length $b = 8$ in. The web thickness $t_w = \frac{3}{8}$ in. for the W21 × 50.

Table V-A: With a length $b = 8$ in. and a web thickness $t_w = \frac{3}{8}$ in., select a $\frac{3}{4}$-in.-thick angle that has a capacity of 31.2 k.

Table V-C: For $\frac{7}{8}$-in. A325-N bolts select a type B connection with a capacity of 50.5 k. (A type D connection with three rows of bolts cannot be used because the web of the column will interfere with the bolt in the middle row.)

Table V-D: For a type B connection an L6 × 4 × $\frac{3}{4}$ is available.

Use four $\frac{7}{8}$-in. A325-N bolts and seat angle L6 × 4 × $\frac{3}{4}$ × 0 ft-8 in.

7-13 STIFFENED SEATED BEAM CONNECTIONS

The load that can be transmitted by a seated beam connection is limited by the bending strength of the outstanding leg of the seat angle. The maximum load capacity listed in the AISC *Manual* tables for a seated connection is 78.4 k. By stiffening the outstanding leg with two vertical angles as shown in Fig. 7-22, the seat can be used to support heavier loads. The moment acting on the bolt group due to the location of the reaction can be neglected in the design provided that the outstanding leg is not too large. From the AISC *Manual* table for stiffened seats it would appear that for outstanding legs 5 in. or less, the effect of the moment on the bolt group can be neglected in the design. The design procedure will be illustrated in the following examples.

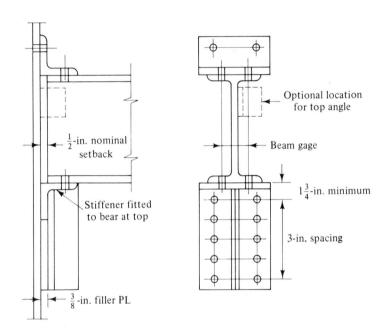

Figure 7-22   Stiffened seated beam connection.

Example 7-14

Design a stiffened beam seat connection for a W27 × 94 beam of A36 steel. The reactive force is $R = 70$ k. The beam seat is connected to the $\frac{1}{2}$-in. web of a column by $\frac{7}{8}$-in. A325-N bolts with a gage of $5\frac{1}{2}$ in.

Solution.   Properties of W27 × 94 beam: $t_w = \frac{1}{2}$ in., $k = 1\frac{7}{16}$ in. The required bearing length N is determined from the stress in the web of the beam and is found from Eq. (5-9) [AISC Formula 1.10-9]. Solving for N, we have

$$N = \frac{R}{0.75 F_y t} - k = \frac{70}{0.75(36)(0.5)} - 1.438 = 3.74 \text{ in.}$$

If the setback is taken as $\frac{3}{4}$ in., the legs of a seat angle $l = 3.74 + 0.75 = 4^+$ in. are required. Use 5 in. The stiffener angles will then be 4 in. The thickness of the two stiffener angles will be selected so that the bearing stress on the stiffeners is no greater than the allowable bearing stress. Thus

$$f_p = \frac{R}{2lt} \le F_p = 0.9F_y \qquad \text{(AISC Specification 1.5.1.5.1)}$$

Solving for t, we have

$$t \ge \frac{R}{1.8F_y l} = \frac{70}{1.8(36)(4)} = 0.270 \text{ in.} \quad \text{(use } \tfrac{5}{16} \text{ in.)}$$

The capacity in single shear for a $\frac{7}{8}$-in. A325-N bolt is $r_v = 12.6$ k (AISC *Manual*, Table I-D). The bearing capacity for the bolt on a $\frac{3}{8}$-in.-thick seat angle with a spacing of $2\frac{1}{3}$ in. is 20.6 k (AISC *Manual*, Table I-E). Therefore, bearing on the stiffeners or the web of the column will not control. The number of bolts required $N = R/r_v = 70/12.6 = 5^+$ (use 6 bolts).

Check of the design by AISC Manual tables. Table VII-A: With $F_y = 36$ ksi a bearing length of 3.74 in. requires the oustanding leg of the stiffener to be 4 in. With a stiffener thickness $t = \frac{5}{16}$ in. the reaction $R = 70.9$ k $>$ 70 k. (OK) The seat angle is $\frac{3}{8}$ in. thick, and to extend beyond the stiffeners, use a 5-in. leg.

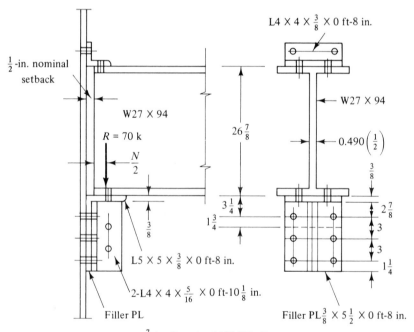

(All dimensions in inches except as noted)

Figure 7-23   Example 7-14.

Table VII-B: With $\frac{7}{8}$-in. A325-N bolts six bolts or three bolts in each vertical row support 75.8 k > 70 k. (OK)   The bearing capacity for six bolts on a thickness of $\frac{3}{8}$ in. with a spacing of 2.33 in. is 6(20.6) = 123.6 k > 70 k (AISC *Manual*, Table I-E). Therefore, bearing on the bolts will not control. The design for the beam seat is shown in Fig. 7-23.

7-14 MOMENT-RESISTING CONNECTIONS

There are two types of moment-resisting connections: Type 1, rigid framing, and Type 3, semirigid framing.

Rigid Connections

Type 1 connections are used in rigid or continuous frame construction. The properly designed rigid connection performs as predicted by maintaining close to a constant angle between the members and by developing a moment near the full moment capacity of the members. The rigid frame can support horizontal loads such as wind or earthquakes without bracing. The rigid frame has a considerable margin of safety against accidental or unexpected overload. Rigid connections are required for frames that are designed by plastic or limit design methods.

Semirigid Connection

Type 2 connections have been used in semicontinuous frame construction for buildings such as offices or apartments of moderate height, where they have proven to be economical. With semirigid connections the center span moments are reduced from those that would occur if the span was simply supported. This can result in a saving in the weight of the beams.

For bolted connections the AISC Specification 1.2 permits the use of simple connections, Type 2, for gravity loads together with semirigid connections, Type 3, at selected joints in the frame to resist the wind loads.

The tee stub with web-angle moment connection shown in Fig. 7-24 can be used as either a rigid or a semirigid connection. As a rigid connection it may be uneconomical since it uses a considerable amount of material and it may be difficult to develop the full moment capacity of the beam.

In the tee stub with web-angle moment connection, the shear force V is transmitted to the column by the two web angles. The bending moment is transmitted to the column flange by the two tees that are attached to the top and bottom flange of the beam. On the compression side of the beam the tee is in compression and exerts a force P_C on the column. The web of the tee and column must be checked for crippling. On the tension side the tee and

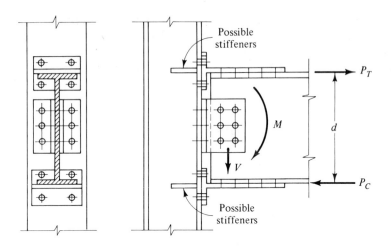

Figure 7-24   T-stub with web angle moment connection.

bolts are in tension and exert a force P_T on the column. The flanges of the tee are subjected to bending stresses and deformations that cause a prying action in the connection, as shown in Fig. 7-25(a). The prying action exerts a prying force at the outside edges of the tee that adds directly to the tension in the bolts, as shown in Fig. 25(b).

In designing the tee stub with web-angle connection we will follow the design method given in the AISC *Manual*, pp. 4-88 through 4-90. The method is illustrated in the following examples.

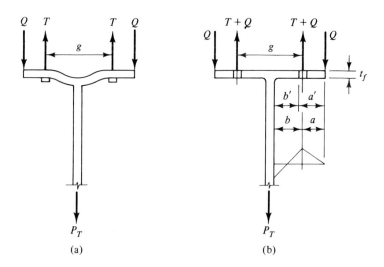

Figure 7-25   Prying action.

Example 7-15

Design tee stubs with web angles to connect a W16 × 40 beam to a W14 × 68 column. Assume a negative moment $M = 85$ k-ft and a shear $V = 37$ k. The member is of A36 steel and the connectors are $\frac{3}{4}$-in. A325-X bolts. A sketch of the connection is shown in Fig. 7-26.

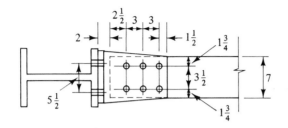

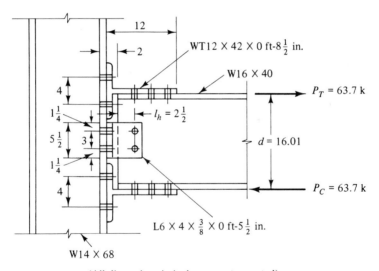

(All dimensions in inches except as noted)

Figure 7-26   Example 7-15.

Solution.   The web angles will be designed to support the entire shear force and the tees to support the bending moment. Properties of the W16 × 40 beam: $d = 16.01$ in., $b_f = 6.995$ in., $t_f = 0.505$ in., and $k = 1\frac{3}{16}$ in.

Design of web angles

Web angle thickness and number of bolts required. Table II-A: For $\frac{3}{4}$-in. A325-X bolts the suggested angle thickness $t = \frac{3}{8}$ in. with $N = 2$ bolts, a length of $l = 5\frac{1}{2}$ in., and a shear capacity of 53.0 k > 37 k.   (OK)

Shear capacity of angles. Table II-C: The shear capacity of the angles if two $\frac{3}{4}$-in. bolts are used is 50.6 k > 37 k. (OK)

Bearing capacity of the web. Table I-E: With $F_u = 58$ ksi, a web thickness $t_w = 0.305$ in., two $\frac{3}{4}$-in. bolts, and the usual 3-in. spacing, the allowable load is $2(0.305)(65.3) = 39.8$ k > 37 k. (OK)

Edge distance. Table I-F: For $F_u = 58$ ksi, $l_h = 2\frac{1}{4}$ in., two $\frac{3}{4}$-in. bolts, and a web thickness $t_w = 0.305$ in., the capacity is $2(0.305)(65.3) = 39.8$ k > 37 k. (OK)

Use two L6 × 4 × $\frac{3}{8}$ × 0 ft-5$\frac{1}{2}$ in.

Design of flange tee connections. From Fig. 7-24

$$P_C = P_T = P \qquad \text{and} \qquad M = Pd$$

Therefore,

$$P = \frac{M}{d} = \frac{85(12)}{16.01} = 63.7 \text{ k}$$

The number of tee flange $\frac{3}{4}$-in. A325-X tension bolts required with $r_t = 19.4$ k (AISC *Manual*, Table I-A) is

$$N = \frac{P}{r_t} = \frac{63.7}{19.4} = 3^+ \quad \text{(use 4 minimum)}$$

The force on each of four bolts is

$$T_b = \frac{P}{N} = \frac{63.7}{4} = 15.92 \text{ k}$$

The number of tee stem bolts in single shear required with $r_v = 13.3$ k (AISC

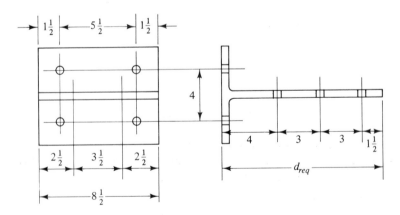

(All dimensions in inches)

Figure 7-27   Example 7-15.

Manual, Table I-D) is

$$N = \frac{P}{r_v} = \frac{63.7}{13.3} = 4^+ \quad \text{(use 6 minimum)}$$

Try flange bolts and web bolts with gages as shown in Fig. 7-27 for the flange tee connection. For the tee stub try WT12 × 47: $t_w = 0.515$ in., $b_f = 9.065$ in., $t_f = 0.875$ in., $d = 12.155$ in. $> d_{\text{req}} = 11.5$ in. The tensile capacity of the tee web based on the yield of the gross section

$$P_t = 0.6F_ybt = 22.0(8.5)(0.515)$$

$$= 96.3 \text{ k} > 63.7 \text{ k} \qquad \text{OK}$$

and the failure of the net effective section

$$P_t = 0.5F_u\left[w - n\left(d + \frac{1}{8}\right)\right]t_w$$

$$= 0.5(58)\left[8.5 - 2\left(\frac{3}{4} + \frac{1}{8}\right)\right](0.515)$$

$$= 100.8 \text{ k} > 63.7 \text{ k} \qquad \text{OK}$$

Check of tee flange for tension and prying. See AISC *Manual,* p. 4-89, for equations and definition of symbols.

$$p = \frac{8.5}{2} = 4.25 \text{ in.}$$

$$b = \frac{g - t_w}{2} = \frac{4.0 - 0.515}{2} = 1.742 \text{ in.}$$

$$b' = b - \frac{d}{2} = 1.742 - \frac{0.75}{2} = 1.367 \text{ in.}$$

$$a = \frac{b_f - g}{2} = \frac{9.065 - 4.0}{2} = 2.532 \text{ in.}$$

$$a \le 1.25b = 1.25(1.742) = 2.178 \text{ in.}$$

$$a' = a + \frac{d}{2} = 2.178 + \frac{0.75}{2} = 2.553 \text{ in.}$$

$$d' = d + \frac{1}{16} = \frac{3}{4} + \frac{1}{16} = 0.8125 \text{ in.}$$

$$\delta = 1 - \frac{d'}{p} = 1 - \frac{0.8125}{4.25} = 0.809 \text{ in.}$$

$$M = \frac{pt_f^2 F_y}{8} = \frac{4.25(0.875)^2(36)}{8} = 14.64 \text{ k-in.}$$

$$\alpha = \frac{\left(\dfrac{Tb'}{M}\right) - 1}{\delta} = \frac{\dfrac{15.92(1.367)}{14.64} - 1}{0.809} = 0.601$$

$$B_c = T\left[1 + \frac{\delta\alpha}{(1 + \delta\alpha)}\left(\frac{b'}{a'}\right)\right]$$

$$= 15.92\left[1 + \frac{0.809(0.601)}{1 + 0.809(0.601)}\left(\frac{1.367}{2.553}\right)\right]$$

$$= 18.71\,\text{k} < 19.4\,\text{k} \qquad \text{OK}$$

$$t_{f\min} = \left[\frac{8B_c a'b'}{pF_y[a' + \delta\alpha(a' + b')]}\right]^{1/2}$$

$$= \left[\frac{8(18.71)(2.553)(1.367)}{4.25(36)[2.553 + 0.809(0.601)(2.553 + 1.367)]}\right]^{1/2}$$

$$= 0.875\,\text{in.} = 0.875\,\text{in.} \qquad \text{OK}$$

$$Q = B_c - T = 18.71 - 15.92 = 2.79\,\text{k}$$

Use a WT12 × 42 − $8\frac{1}{2}$-in. long.

Check for possible column web stiffeners. Properties of W14 × 68 column: $d = 14.04$ in., $t_w = 0.415$ in., $b_f = 10.035$ in., $t_f = 0.720$ in., $k = 1.5$ in. From AISC Specification 1.15.5.2,

$$P_{bf} = \frac{5}{3}P = \frac{5}{3}(63.7) = 106.2\,\text{k}$$

From AISC Formula (1.15-1),

$$A_{st} = \frac{P_{bf} - F_{yc}t(t_b + 5k + 2t_{fwt})}{F_{yst}}$$

$$= \frac{106.2 - 36(0.415)[0.515 + 5(1.5) + 2(0.875)]}{36}$$

$$= -1.102\,\text{in.}^2$$

The term $2t_{fwt}$ has been added to account for the additional spread of forces at 45° through the flange of the tee section.

From AISC Formula (1.15-2) for the compression tee connection,

$$d_c = d - 2k = 14.04 - 2(1.5) = 11.04\,\text{in.}$$

$$d_c = \frac{4100t^3\sqrt{F_{yc}}}{P_{bf}} = \frac{4100(0.415)^3\sqrt{36}}{106.2} = 16.6\,\text{in.}$$

$$11.05\,\text{in.} < 16.6\,\text{in.} \qquad \text{OK}$$

From AISC Formula (1.15-3) for the tension tee connection,

$$t_f \geq 0.4 \sqrt{\frac{P_{bf}}{F_{yc}}} = 0.4 \sqrt{\frac{106.2}{36}} = 0.687 \text{ in.}$$

$$0.720 \text{ in.} > 0.687 \text{ in.} \qquad \text{OK}$$

No stiffeners are required.

See Example 8-22 and Moment Connections, AISC *Manual*, pp. 4–98 through 4–110, for the design of web stiffeners.

PROBLEMS

Use the AISC *Specifications* for the following problems. Assume that all holes are standard and the surface condition for the bolted parts is clean mill scale (Classification A) unless indicated otherwise.

7-1. Two $\frac{1}{2}$-in. × 10-in. plates of A36 steel are joined by $\frac{7}{8}$-in.-diameter bolts or rivets as shown. Determine the tensile capacity of the plates if the bolts or rivets are as follows:

    (a) A502 Grade 1 rivets.

    (b) A325-F bolts.

    (c) A325-N bolts.

    (d) A325-X bolts.

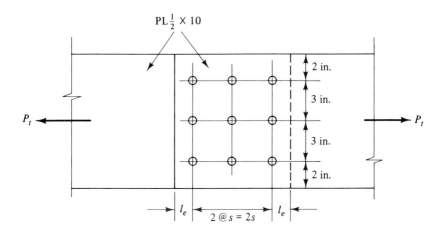

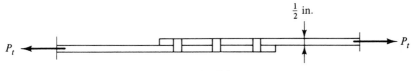

Prob. 7-1

Indicate the minimum values for connector spacing s and edge distance l_e for the connection.

7-2. Repeat Prob. 7-1 using 1-in.-diameter bolts or rivets.

7-3. Determine the number of $\frac{3}{4}$-in.-diameter bolts required in the butt splice shown to develop the full tensile capacity P_t of the member if it is made of A36 steel.

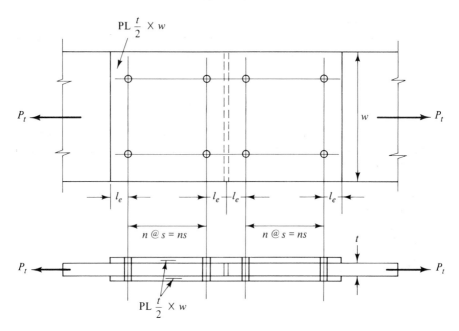

Prob. 7-3

The width w, thickness t, number of bolts n in a cross section, and type of bolt are as follows:

(a) $w = 12$ in., $t = \frac{3}{4}$ in., $n = 3$, A325-X.

(b) $w = 8$ in., $t = \frac{5}{8}$ in., $n = 3$, A325-X.

(c) $w = 6$ in., $t = \frac{1}{2}$ in., $n = 2$, A325-F.

(d) $w = 6$ in., $t = \frac{1}{2}$ in., $n = 2$, A325-X.

Indicate the minimum values for bolt spacing s and edge distance l_e for the connection.

7-4. Repeat Prob. 7-3. Use A572 Grade 50 steel.

7-5. Determine the number of bolts required to develop the full tensile capacity P_t of the double-angle tension member shown. Use A36 steel. There are two rows of bolts without stagger and the long legs are back to back. The angle size and type of bolt are as follows:

(a) Two L8 × 6 × $\frac{1}{2}$, $\frac{7}{8}$-in.-diameter A325-F.

(b) Two L8 × 6 × $\frac{1}{2}$, $\frac{7}{8}$-in.-diameter A325-X.

(c) Two L6 × 4 × $\frac{3}{8}$, $\frac{3}{4}$-in.-diameter A325-F.

(d) Two L6 × 4 × $\frac{3}{8}$, $\frac{3}{4}$-in.-diameter A325-N.

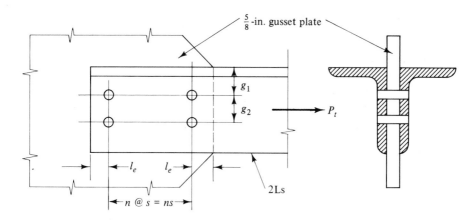

Prob. 7-5

Indicate the minimum value for bolt spacing s and edge distance l_e for the connection.

7-6. Repeat Prob. 7-5. Use A441 steel with $F_y = 50$ ksi.

7-7. A tension member of A36 steel consists of two C10 × 20 connected to a $\frac{7}{8}$-in. gusset plate as shown. Determine the number of bolts required to develop the full tensile capacity of the channels if there are four $\frac{3}{4}$-in.-diameter bolt holes in

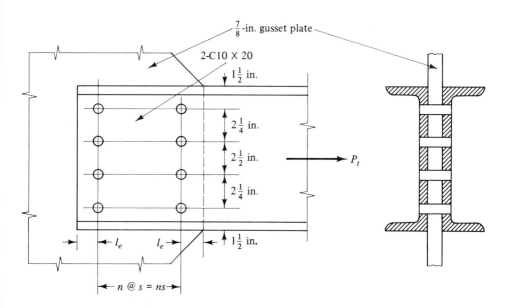

Prob. 7-7

a cross section and the bolts are as follows:

(a) A325-F.

(b) A325-X.

(c) A490-F.

(d) A490-X.

Indicate the minimum bolt spacing s and edge distance l_e for the connection.

7-8. Repeat Prob. 7-7 if there are three $\frac{7}{8}$-in.-diameter bolt holes in a cross section.

7-9. Determine the maximum force on the bolts in the bolt group shown. Use

(a) Elastic analysis.

(b) Ultimate strength analysis (AISC *Manual*).

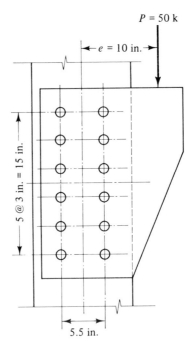

$P = 50$ k

$e = 10$ in.

5 @ 3 in. = 15 in.

5.5 in.

Prob. 7-9

7-10. Repeat Prob. 7-9 witth $P = 75$ k and $e = 11.5$ in.

7-11. Determine the number of $\frac{7}{8}$-in.-diameter bolts required for the eccentrically loaded bracket shown. Use A36 steel. The load P, eccentricity e, and type of bolt are as foliows:

(a) $P = 69$ k, $e = 8$ in., A325-F.

(b) $P = 98$ k, $e = 10$ in., A325-N.

(c) $P = 140$ k, $e = 10$ in., A325-X.

(d) $P = 163$ k, $e = 12$ in., A325-X.

7-12. Repeat Prob. 7-11 if the loads are reduced by 25 percent and $\frac{3}{4}$-in.-diameter bolts are used.

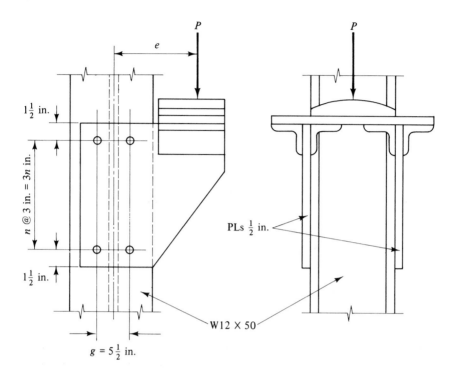

Prob. 7-11

7-13. Determine if the connection shown in Fig. 7-16 is adequate when $P_t = 80$ k and $\theta = 30°$ if six
(a) $\frac{3}{4}$-in.-diameter A325-N bolts are used.
(b) $\frac{7}{8}$-in.-diameter A325-F bolts are used.
Disregard the effect of prying.

7-14. Repeat Prob. 7-13 for $P_t = 65$ k and $\theta = 55°$.

7-15. Determine the number of bolts required for connection A and B for the bracket shown. Use A36 steel. The load P_t, slope angle θ, bolt diameter d, and type of bolt for connections A and B are as follows:
(a) 120 k, 30°, $\frac{7}{8}$ in., A325-N, A490-F.
(b) 120 k, 30°, $\frac{7}{8}$ in., A325-X, A490-X.
(c) 85 k, 35°, $\frac{3}{4}$ in., A325-N, A490-F.
(d) 85 k, 45°, $\frac{3}{4}$ in., A325-X, A490-X.
Indicate the minimum values for connector spacing s and edge distance l_e for connections A and B.

7-16. Repeat Prob. 7-15 using A572 Grade 50 steel.

7-17. Determine the capacity for the double-angle web-framing connection shown if a friction-type connection is used.

7-18. Repeat Prob. 7-17 as a bearing-type connection with the threads excluded from the shear plane.

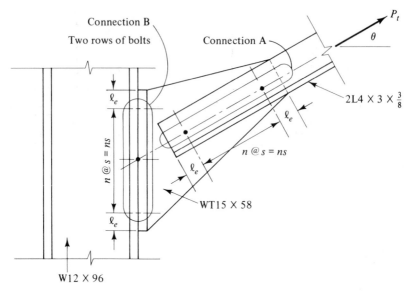

Prob. 7-15

7-19. Determine the number of bolts required to connect a W18 × 40 beam to the flange of a W12 × 65 column with a double-angle web-framing connection and the minimum horizontal edge distance l_h for the beam. Rolled sections are of A36 steel. The reactive force to be transmitted is 58 k. Use $\frac{3}{4}$-in.-diameter A325-N bolts.

7-20. Repeat Prob. 7-19 as a friction-type connection.

7-21. Determine the number of bolts required to connect a W21 × 50 beam to a W36 × 135 girder with a double-angle web-framing connection. Rolled sections

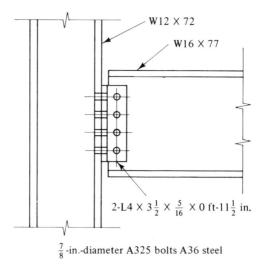

$\frac{7}{8}$-in.-diameter A325 bolts A36 steel   **Prob. 7-17**

are of A36 steel. The reactive force to be transmitted is $R = 79$ k. The flange of the beam will be coped. Use $\frac{3}{4}$-in.-diameter A325-N bolts.

7-22. Design a double-angle web-framing connection for the W18 × 106 beam shown. The end reaction is 80 k and all members are of A36 steel. Use $\frac{7}{8}$-in. A325-X bolts.

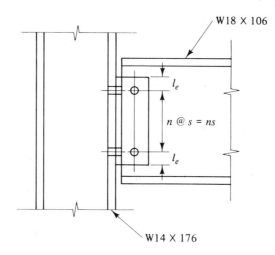

Prob. 7-22

7-23. Is the connection shown in the figure satisfactory? The base metal is A36 steel. The top eight bolts each receive one-eighth of the reaction of the beam on the left and one-twelfth of the reaction of the beam on the right.

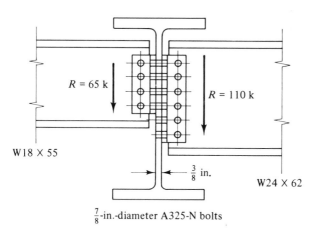

$\frac{7}{8}$-in.-diameter A325-N bolts

Prob. 7-23

7-24. Design web angle connections for a W16 × 40 beam having a reaction of $R = 50$ k and a W21 × 50 beam with a reaction of $R = 75$ k framing into opposite sides of the web of a W24 × 84 girder in the same manner as shown in the figure for Prob. 7-23. The members are made of A36 steel. Use $\frac{3}{4}$-in.-diameter A325-X bolts.

7-25. Design a seated beam connection to support a W18 × 55 beam of A36 steel. The reactive force $R = 28$ k. The beam seat is connected to a W12 × 50 column web with a gage of $5\frac{1}{2}$-in. by $\frac{7}{8}$-in.-diameter A325-F bolts.

7-26. Repeat Prob. 7-25 as a bearing-type connection with the threads excluded from the shear plane.

7-27. Design a seated beam connection to support a W21 × 83 beam of A36 steel. The reactive force is 40 k. The beam seat is connected to a W12 × 79 column flange by $\frac{7}{8}$-in.-diameter A325-N bolts with a gage of $5\frac{1}{2}$ in.

7-28. Repeat Prob. 7-27 as a friction-type connection.

7-29. Design a stiffened beam seat connection for a W24 × 84 beam of A36 steel. The reactive force is $R = 62.5$ k. The beam seat is connected to the web of a W12 × 87 column with a gage of $5\frac{1}{2}$ in. Use $\frac{7}{8}$-in.-diameter A325-F bolts.

7-30. Repeat Prob. 7-29 as a bearing-type connection with threads included in the shear plane.

7-31. Design a tee stub with web angle connection as shown in Fig. 7-24 to connect a W18 × 40 beam to a W14 × 74 column. Assume a negative moment $M = 110$ k-ft and a shear $V = 40$ k. The members are connected by $\frac{3}{4}$-in.-diameter A325-N bolts. Assume that forces are due to combined dead, live, and wind loads. Try a WT12 for the tee.

7-32. Repeat Prob. 7-31 with $\frac{7}{8}$-in.-diameter A325-X bolts. Try a WT9 for the tee.

8 WELDED CONNECTIONS

8-1 INTRODUCTION

Welding is the joining of two pieces of metal by creating a strong metallurgical bond between them by heat or pressure or both. There are numerous welding processes, but the one most commonly used in civil engineering structures is *electric-arc welding.* In this process the base metal and welding rod are heated to the fusion temperature by an electric arc. The electric arc is formed when a large current at low voltage discharges between an insulated welding rod and the object to be welded through a thermally ionized gaseous column. The welding rod is connected to one terminal of the current source and the object to be welded to the other. Fusion takes place by the flow of material from the welding rod across the arc. No pressure is applied.

Arc Welding Processes

In *shielded-metal arc welding* or the SMAW process [Fig. 8-1(a)] the electrode coating produces a gaseous "shield" that helps to exclude oxygen and stabilizes the arc. The coating also produces slag that protects the weld from the atmosphere and slows cooling. After cooling, the slag can be removed by preening and a wire brush.

The arc is "submerged" or covered by a mound of fusible powdered flux in the *submerged-arc welding* or SAW process and the bare electrode wire is fed mechanically from a reel. The arc is at all times covered by the flux as illustrated in Fig. 8-1(b). The heat of the arc melts the electrode, the object

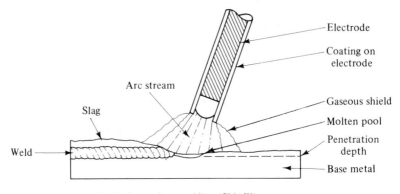

(a) Shielded-metal arc welding (SMAW) process

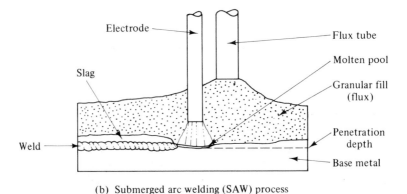

(b) Submerged arc welding (SAW) process

Figure 8-1   Welding processes.

to be welded, and part of the flux. The flux forms a slag covering that can be removed later by brushing. The greater heat input in the SAW process results in deeper penetration than for the SMAW process. This is accounted for in design.

Another welding process of interest is *gas-shielded arc welding* or the GMAW process. It is used mostly in shop welding. An inert gas provides the shield against the atmosphere for the uncoated electrode that is used in a mechanical welding machine.

Electroslag welding or the ESW process is similar to submerged-arc welding. However, it uses an electroconductive slag. The slag is held in position by water-cooled retaining plates. The process is started by an electric arc, but when the molten slag bath forms, the arc is extinguished. The current continues to flow through the electroconductive slag. Thus the ESW process is not a true arc welding process. It is most often used to join thick plates together in a vertical position.

Advantages of Welding

Welding is permitted for most structural steel fabrication. Welded connections are used for most shop connections and a considerable number of field connections. They offer many advantages over bolts and rivets. There is considerable saving in weight because of simpler connections and the use of members without a reduction for holes. Because of the elimination of punching and drilling holes and assembling for fit, fabrication costs and time are reduced. Welded connections are practical for curved members such as structural pipes. Welding lends itself to smooth, clean lines that enhance the appearance of the structure and reduce stress concentrations due to local irregularities.

Good welded connections are not just imitations of bolted connections with welds substituted for bolts. With careful design and fabrication the disadvantages of welding such as fatigue and cracking due to the rigidity of the connections can be minimized or eliminated.

8-2 TYPES OF JOINTS AND WELDS

A number of different types of basic structural joints and welds are illustrated in Fig. 8-2. For lap joints the ends of two members are overlapped, and for butt joints the two members are placed end to end. The T-joint forms a tee and for corner joints the ends are joined like the letter L. Edge joints are usually not structural joints. They can be used to maintain alignment of the edges of two or more plates.

The most common type of welds are the *fillet weld* and the *groove weld.* Fillet welds can be used for lap joints and T-joints and groove welds for butt joints and corner joints. When groove welds are made from both sides or from one side with a backup strip on the far side, they attain complete penetration and are called *complete-penetration groove welds.* They may be stressed as fully as the weakest part being joined. *Incomplete-penetration groove welds* are used where full continuity is not required and the parts are not fully stressed. A description of a welded joint requires an indication of the type of both joint and weld. For example, the T-joint could be either fillet or groove welded. Various prequalified welded joints are illustrated in the AISC *Manual,* pp. 4-149 through 4-165.

8-3 WELD SYMBOLS

Standard weld symbols used in steel detailing are given in the AISC *Manual,* p. 4-148. The weld symbols of the AISC *Specifications* parallel the recommendation of the American Welding Society (AWS).

To help clarify the use of the symbols, several examples together with an explanation are given in Fig. 8-3.

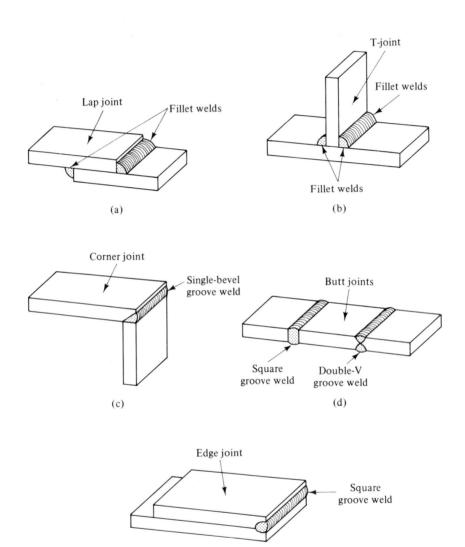

Figure 8-2 Types of joints and welds.

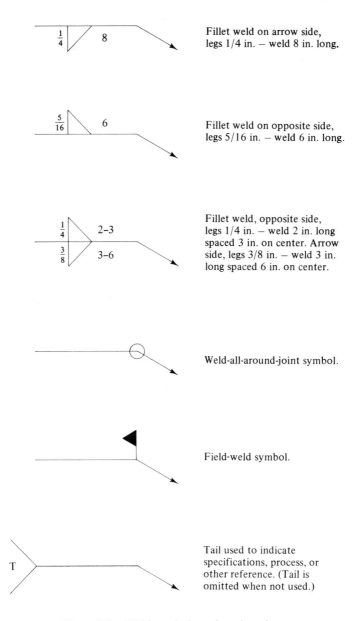

Figure 8-3 Weld symbols and explanations.

8-4 ALLOWABLE WELD STRESSES

When an electrode has properties that are compatible with the base metal, the electrode is described as "matching" the base metal. Current (1978) AISC *Specifications* refer to Table 4.1.1, "Structural Welding Code," AWS D1.1-77 (or latest edition) for various matching grades of base metal and electrodes. Some of this information is summarized in Table 8-1.

TABLE 8-1   Matching Filler Metal Requirements[a]

Type of Base Metal ASME Designation	Welding Process	
	Shielded Metal Arc Welding (SMAW)	Submerged Arc Welding (SAW)
A36, A53 Grade B, A500 Grades A and B, A501, A529	E60XX or E70XX[b]	F6X-EXXX or F7X-EXXX[c]
A242, A441, A572 Grades 42 and 50, A588 (4 in. and under)	E70XX	F7X-EXXX
A572 Grades 60 and 65	E80XX	F8X-EXXX
A514 (over $2\frac{1}{2}$ in. thick)	E100XX	F10X-EXXX
A514 ($2\frac{1}{2}$ in. and under)	E110XX	F11X-EXXX

[a]Consult the latest AWS Code for detailed information.
[b]The E stands for electrode; the first two or three digits indicate minimum tensile strength in ksi; and the X's, when replaced by the appropriate digit, indicate welding position, current, and other welding procedure variables.
[c]The F indicates granular flux; the first one or two digits when multiplied by 10 indicate the tensile strength in ksi; the X, when replaced by the appropriate digit and multiplied by -10, indicates the testing temperature in °F for weld metal impact tests; the E stands for electrode; and the last three X's, when replaced by the appropriate letters and digits, indicate the electrode used.
Source: Abridged from Table 4.1.1 of the "Structural Welding Code," AWS D1.1-84. Reproduced by permission of the American Welding Society.

Allowable stress for complete- and partial-penetration groove welds and fillet welds are given in Table 1.5.3 of the AISC *Specifications* and summarized in Table 8-2. The allowable stress is applied to the effective area of a weld. According to AISC Specification 1.14.6, the effective area for groove and fillet welds is the effective length multiplied by the effective throat thickness.

TABLE 8-2   **Allowable Weld Stresses**

Type of Weld	Type of Stress	Allowable Stress
Complete-penetration groove weld	Compression parallel or normal to weld axis and tension parallel to weld axis	Same as base metal[a]
	Tension normal to effective area	Same as base metal[b]
Partial-penetration groove weld	Compression parallel or normal to weld axis and tension parallel to weld axis	Same as base metal[a]
	Tension normal to effective area	$0.3F_u$ (electrode)[a] except tensile stress on base metal not to exceed $0.4F_y$ (base metal)
All groove welds	Shear	$0.3F_u$ (electrode)[a] except shear stress on base metal not to exceed $0.4F_y$ (base metal)
Fillet welds	Shear	$0.3F_u$ (electrode)[a] except shear stress on base metal not to exceed $0.4F_y$ (base metal)

[a]Weld metal with a strength level equal to or less than "matching" weld metal may be used.
[b]"Matching" weld metal must be used.
Source: Adapted from AISC *Specifications*, Table 1.5.3. (Courtesy of the American Institute of Steel Construction, Inc.)

Groove Welds

The *effective length* for a groove weld is the width of the plates joined. The *effective throat thickness* for the complete-penetration groove weld is the thickness of the thinnest part joined. The effective throat for a partial-penetration groove weld depends on the welding process, welding position, and angle at the root of the groove. Thicknesses are given in Table 1.14.6.1.2 of the AISC *Specifications*.

Fillet Welds

The effective length, except in holes and slots, is the overall length of the full-size fillet, including returns. The theoretical throat t and effective throat T_e are illustrated in Fig. 8-4. If the weld has equal legs of nominal size a, the theoretical throat

$$t = 0.707a \qquad\qquad (a)$$

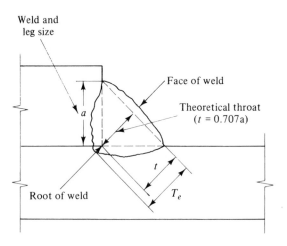

Figure 8-4   Fillet weld terminology.

With the manual shielded-metal arc welding (SMAW) process, the effective throat is taken equal to the theoretical throat. Thus

$$T_e = t \qquad\qquad (b)$$

For the submerged metal arc welding (SAW) process, greater heat input is used. This causes deeper penetration and a larger effective throat. When the weld size a is $\frac{3}{8}$ in. or less, the theoretical throat

$$T_e = a \qquad (a \le \tfrac{3}{8}\text{ in.}) \qquad\qquad (c)$$

and when a is greater than $\frac{3}{8}$ in.,

$$T_e = t + 0.11 \qquad (a > \tfrac{3}{8}\text{ in.}) \qquad\qquad (d)$$

The shear strength for a fillet weld per inch of weld is given by

$$q = T_e F_v \qquad\qquad (e)$$

where T_e is the effective throat and F_v is the allowable shear stress in the weld. From Table 8-2 the allowable shear stress F_v is equal to 0.3 multiplied by the tensile strength of the electrode, that is, $F_v = 0.3 F_u$ (electrode).

For the SMAW process and a weld size $D/16$ in., the allowable strength from Eqs. (a), (b), and (e) is

$$q = T_e F_v = t F_v$$

$$= 0.707 a F_v = 0.707 \frac{D}{16} F_v$$

For an electrode with a tensile strength $F_u = 60$ ksi, the allowable shear stress $F_v = 0.3 F_u = 0.3(60) = 18$ ksi. Therefore,

$$q = 0.707 \frac{D}{16} 18 = 0.80 D$$

For the SAW process and a weld size $D/16$ in. with D equal to or less than 6, the allowable shear strength from Eqs. (c) and (e) is

$$q = T_e F_v = a F_v$$

$$= \frac{D}{16} F_v$$

For an electrode with a tensile strength $F_u = 70$ ksi, the allowable shear stress $F_v = 0.3 F_u = 0.3(70) = 21$ ksi. Therefore,

$$q = \frac{D}{16} 21 = 1.31 D$$

For the SAW process and a weld size $D/16$ in. with D greater than 6, the allowable shear strength from Eqs. (a), (d), and (e)

$$q = T_e F_v = (t + 0.11) F_v$$

$$= (0.707 a + 0.11) F_v = \left(0.707 \frac{D}{16} + 0.11 \right) F_v$$

For an electrode with a tensile strength $F_u = 80$ ksi, the allowable shear stress $F_v = 0.3 F_u = 0.3(80) = 24$ ksi. Therefore,

$$q = \left(0.707 \frac{D}{16} + 0.11 \right) 24 = 1.06 D + 2.64$$

The shear strength for various weld processes and electrodes is summarized in Table 8-3.

TABLE 8-3   Shear Strength q_y for Fillet Welds, Weld Size $D/16$ in.[a]

F_u (ksi)	60	70	80	90	100	110
$F_v = 0.3F_u$ (ksi)	18.0	21.0	27.0	27.0	30.0	33.0
SMAW all sizes	$0.80D$	$0.93D$	$1.06D$	$1.19D$	$1.33D$	$1.46D$
SAW $a \leq \frac{3}{8}$ in.	$1.12D$	$1.31D$	$1.50D$	$1.69D$	$1.88D$	$2.06D$
$a > \frac{3}{8}$ in.	$0.80D$ $+1.98$	$0.93D$ $+2.31$	$1.06D$ $+2.64$	$1.19D$ $+2.97$	$1.33D$ $+3.30$	$1.46D$ $+3.63$

[a]Capacity given in kips per inch of weld.

8-5 LIMITATIONS ON SIZE AND LENGTH OF WELDS

Minimum Weld Size

Heating and cooling during the welding process causes the formation of residual stresses in the weld and changes in the metallurgical properties of the base metal. To minimize both of these effects, a minimum weld size is given in AISC Specification 1.17.2 based on the thickness of the thicker of the two base metal parts being joined. These requirements are given in Table 8-4.

TABLE 8-4   Minimum-Size Fillet Weld

Thickness of Thicker Base Metal Plate Being Joined, t (in.)	Minimum[a] Size of Fillet Weld (in.)
$t \leq \frac{1}{4}$	$\frac{1}{8}$
$\frac{1}{4} < t \leq \frac{1}{2}$	$\frac{3}{16}$
$\frac{1}{2} < t \leq \frac{3}{4}$	$\frac{1}{4}$
$t > \frac{3}{4}$	$\frac{5}{16}$

[a]Leg dimension of fillet welds.
Source: AISC *Specifications*, Table 1.17.2. (Courtesy of the American Institute of Steel Construction, Inc.)

Maximum Fillet Weld Size

The maximum size of fillet weld that may be used along the edge or connected part is limited to ensure full throat thickness. AISC Specification 1.17.3 gives the following limitations:

1. Along edges of connected parts less than $\frac{1}{4}$ in. thick the weld size may be no greater than the thickness of the material.
2. Along edges of the connected parts $\frac{1}{4}$ in. or more thick, the weld size may be no greater than $\frac{1}{16}$ in. less than the thickness of the material unless the weld is built up to obtain full throat thickness.

The requirements are illustrated in Fig. 8-5.

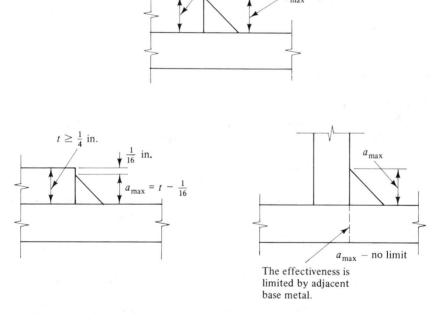

Figure 8-5   Maximum-size fillet weld along edge of connected part (AISC Specification 1.17.3).

Minimum Lengths for Fillet Welds

From AISC Specifications 1.17.4 and 1.17.5 the minimum effective length of fillet welds permitted is four times the weld size or else the weld size is considered to be one-fourth of its effective length, except that for intermittent welds the effective length for any segment must be no less than $1\frac{1}{2}$ in.

When longitudinal fillet welds are used alone in an end connection, the length of each weld must be no less than the perpendicular distance between them. If the transverse spacing of the longitudinal welds exceeds 8 in., provision must be made to prevent excessive transverse bending in the connection.

Minimum Lap for Lap Joints

The minimum amount of lap for a lap joint must be five times the thickness of the thinnest plate joined but not less than 1 in. (AISC Specification 1.17.6).

End Return

Whenever practical, an end return must be provided for a distance not less than two times the weld size. The end return shall be indicated on the design and detail drawing and added to the effective length (AISC Specifications 1.17.7 and 1.14.6.2).

8-6 WELDED CONNECTIONS

The welded connection is inherently continuous and more rigid than the bolted connection. The connection design process begins with the design followed by the welding operation and ends with an inspection.

Connection Design

In connection design due consideration must be taken of the force or stress flow through the connection. The stress flow should occur with the least disturbance possible, to avoid stress concentrations and possible formation of cracks. For example, butt welds made with a smooth transition between the plates have minimal stress concentrations. However, when the weld is made with additional reinforcement that produces an abrupt transition, stress concentrations can be formed that produce stresses substantially larger than the average or nominal value. For members with butt welds that are subject to repeated or vibratory loads the welds can be reinforced by added weld material and ground flush to minimize stress concentrations.

The rigidity of the connection must also be considered. The strains in the various joined parts of the connection must be compatible. That is, parts of the connection joined by welds must have strains that can exist together. This may lead to difficulty when members are framed at right angles to each other. For example, the flanges of a beam welded to the web of a girder must be able to accommodate the transverse strain produced in the flanges due to the bending of the girder.

Welding Operation and Inspection

Edge preparation, type and diameter of electrode, welding current output, welding rate, rate of cooling, and the sequence used in welding a joint or joints in the connection are all factors that affect the quality of a completed connection.

The key to success in welding is inspection and control. The welding industry has formulated rules for welders and welding that have helped to ensure sound welds and reliable welded connections.

8-7 AXIALLY LOADED WELDS

The design of the welded connection for tension or compression members requires the welds to be distributed so that the connection does not introduce eccentricity into the loading and to be at least as strong as the members joined.

Example 8-1

Two $PL\frac{1}{2} \times 6$ of A36 steel are joined by a lap joint as shown. Determine the minimum lap required to develop the full tensile capacity of the plates using

(a) The SMAW process and E60 electrodes.

(b) The SAW process and F7X flux electrode combination.

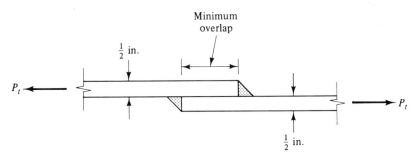

Figure 8-6   Example 8-1.

Solution.   The minimum lap

$$L = 5t \text{ (thinnest plate)} = 5\left(\frac{1}{2}\right) = 2\tfrac{1}{2} \text{ in.} \qquad \text{(AISC Specification 1.17.6)}$$

The tensile capacity of the plate

$$P_t = F_t A_g = 0.6F_y A_g = 22\left(\frac{1}{2}\right)(6) = 66 \text{ k}$$

The maximum weld size

$$a = t - \frac{1}{16} = \frac{1}{2} - \frac{1}{16} = \frac{7}{16} \quad \text{(AISC Specification 1.17.3)}$$

and the minimum weld size

$$a = \tfrac{3}{16} \text{ in.} \quad \text{(AISC Specification 1.17.2)}$$

The effective weld length

$$L = 2(6) = 12 \text{ in.}$$

The required shear strength of the fillet weld

$$q_v = \frac{P_t}{L} = \frac{66}{12} = 5.5 \text{ k/in.}$$

(a) For the SMAW process and E60 electrode ($F_u = 60$ ksi) from Table 8-3: $q_v = 0.80D$. Thus

$$0.80D = 5.5 \qquad D = 6^+ \quad \text{(use 7)}$$

Use $\tfrac{7}{16}$-in. fillet welds.

(b) For the SAW process and F7X flux electrode combination ($F_u = 70$ ksi) from Table 8-3: $q_v = 1.31D$. Thus

$$1.31D = 5.5 \qquad D = 4^+ \quad \text{(use 5)}$$

Use $\tfrac{5}{16}$-in. fillet welds.

Example 8-2

Two PL$t \times 5$ of A36 steel are joined together by a groove weld. The plates support a tensile load of 53 k. Determine the required plate thickness and from Table 8-1 the matching electrodes. Use the SAW process and determine the prequalified complete-penetration square-groove weld required to join the plates.

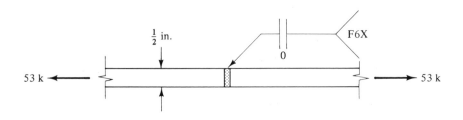

Figure 8-7 Example 8-2.

Solution. The required plate thickness

$$t \geq \frac{P_t}{5(0.6F_y)} = \frac{53}{5(22)} = 0.482 \text{ in.} \quad (\text{use } \tfrac{1}{2} \text{ in.})$$

From Table 8-1 use an F6X flux electrode combination ($F_u = 60$ ksi). See AISC *Manual*, p. 4-150, for prequalified complete-penetration square-groove weld with joint designation B-L1-S. The weld is shown in Fig. 8-7.

Example 8-3

A PL$t \times 7$ of A36 steel is fillet welded to a gusset plate as shown in Fig. 8-8. The plate supports a tensile load of 93 k. Determine the required plate thickness and required SAW process weld lengths if the weld size and flux electrode combination are as follows:

(**a**) Largest efficient weld size, F6X.

(**b**) Smallest possible weld size, F7X.

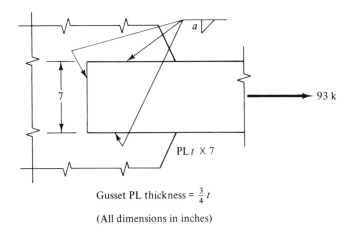

Gusset PL thickness $= \frac{3}{4} t$

(All dimensions in inches)

Figure 8-8   Example 8-3.

Solution. The required plate thickness

$$t \geq \frac{P_t}{7(0.6F_y)} = \frac{93}{7(22)} = 0.603 \text{ in.} \quad (\text{use } \tfrac{5}{8} \text{ in.})$$

The gusset plate thickness

$$t = \frac{3}{4} t = \frac{3}{4}\left(\frac{5}{8}\right) = 0.469 \text{ in.} \quad (\text{use } \tfrac{1}{2} \text{ in.})$$

(a) The largest possible weld

$$a = t - \frac{1}{16} = \frac{5}{8} - \frac{1}{16} = \frac{9}{16} \text{ in.} \qquad \text{(AISC Specification 1.17.3)}$$

The shear capacity of the gusset plate

$$q_v = t(\text{gusset plate}) F_v = t0.4F_y$$

$$= 0.5(0.4)(36) = 7.20 \text{ k/in.} \quad \text{(control)}$$

For the SAW process and F6X flux electrode combination ($F_u = 60$ ksi) from

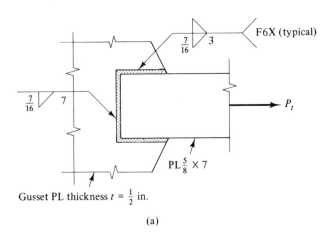

Gusset PL thickness $t = \frac{1}{2}$ in.

(a)

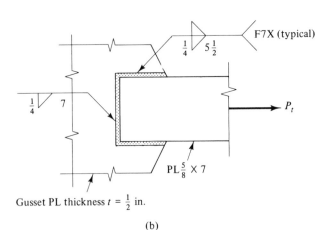

Gusset PL thickness $t = \frac{1}{2}$ in.

(b)

Figure 8-9   Example 8-3.

Table 8-3: $q_v = 1.12 D$. Thus

$$1.12D = 7.20 \qquad D = 6^+ \quad (\text{use } 7)$$

Then

$$q_v = 1.12(7) = 7.84 \text{ k/in.}$$

Use a $\frac{7}{16}$-in. weld.

The length of welds required is determined by the shear capacity of the gusset plate. Therefore,

$$L = \frac{P_t}{q_v} = \frac{93}{7.20} = 12.9 \text{ in.} \quad (\text{use 13 in.})$$

The welds are placed as shown in Fig. 8-9(a).

(b) The smallest weld from Table 8-4 for $t = \frac{5}{8}$ in. is $a = \frac{1}{4}$ in. For the SAW process and F7X flux electrode combination ($F_u = 70$ ksi) from Table 8-3: $D = 4$.

$$q_v = 1.31D = 1.31(4) = 5.24 \text{ k} \quad (\text{control})$$

The shear capacity of the plate does not control. The length of welds required

$$L = \frac{P_t}{q_v} = \frac{93}{5.24} = 17.7 \text{ in.} \quad (\text{use 18 in.})$$

The welds are placed as shown in Fig. 8-9(b).

Example 8-4

Design a balanced fillet welded connection to develop the full tensile capacity of L6 × 4 × $\frac{1}{2}$ of A572 Grade 50 steel shown in Fig. 8-10(a). Assume that the gusset plate does not control.

(a) Use the SMAW process.

(b) Omit the end weld and use the SAW process.

Solution.   The tensile capacity of the angle

$$P_t = 0.6F_y A_g = 0.6(50)(4.75) = 142.5 \text{ k}$$

The smallest weld size from Table 8-4 for $t = \frac{1}{2}$ in. is $a = \frac{3}{16}$ in. and the maximum weld size from Fig. 8-5 is $a = t - \frac{1}{16} = \frac{1}{2} - \frac{1}{16} = \frac{7}{16}$ in.

(a) Try a $\frac{3}{8}$-in. fillet weld with an E70 electrode (Table 8-1). From Table 8-3 for $D = 6$,

$$q = 0.93D = 0.93(6) = 5.58 \text{ k/in.} \quad (\text{control})$$

The shear capacity of the angle

$$\max q_v = 0.4F_y t = 0.4(50)\left(\frac{1}{2}\right) = 10 \text{ k/in.}$$

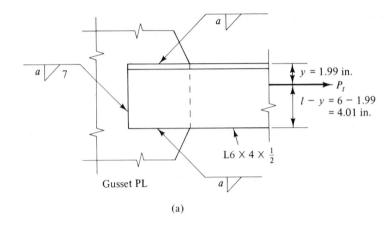

(a)

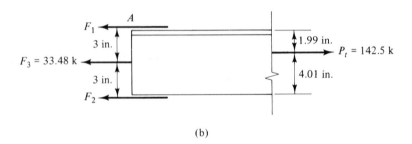

(b)

Figure 8-10   Example 8-4.

The reactive forces for each of the three welds together with the load are
shown in Fig. 8-10(b).

The force $F_3 = qL = 5.58(6) = 33.48$ k. For equilibrium we have from
Fig. 8-10(b)

$$\sum M_A = F_2 6 + 33.48(3) - 142.5(1.99) = 0$$

$$F_2 = 30.5 \text{ k}$$

and

$$\sum F_x = 142.5 - F_2 - F_1 - 33.48 = 0$$

$$F_1 = 78.5 \text{ k}$$

The lengths of welds required

$$L_1 = \frac{F_1}{q} = \frac{78.5}{5.58} = 14.07 \text{ in.} \quad \text{(use 14 in.)}$$

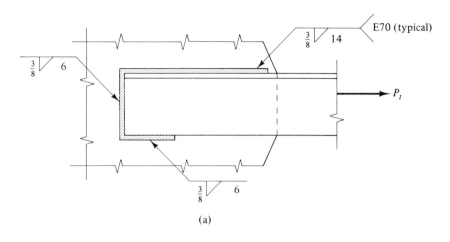

(a)

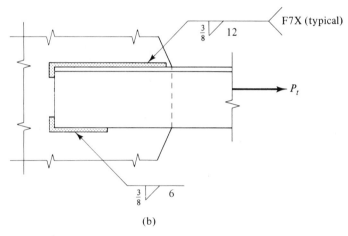

(b)

Figure 8-11   Example 8-4.

and

$$L_2 = \frac{F_2}{q} = \frac{30.5}{5.58} = 5.47 \text{ in.} \quad (\text{use 6 in.})$$

The connection is shown in Fig. 8-11(a).

(**b**) Try a $\frac{3}{8}$-in. fillet weld with an F7X flux electrode combination (Table 8-1). From Table 8-3 with $D = 6$

$$q = 1.31D = 1.31(6) = 7.86 \text{ k} \quad (\text{control})$$

The force $F_3 = 0$ (end weld omitted). For equilibrium we have, from Fig. 8-10(b),

$$\sum M_A = F_2 6 - 142.5(1.99) = 0$$

$$F_2 = 47.3 \text{ k}$$

and

$$\sum F_x = 142.5 - F_2 - F_1 = 0$$

$$F_1 = 95.2 \text{ k}$$

The length of welds required

$$L_1 = \frac{F_1}{q} = \frac{95.2}{7.86} = 12.11 \text{ in.} \quad (\text{use 12 in.})$$

and

$$L_2 = \frac{F_2}{q} = \frac{47.3}{7.86} = 6.02 \text{ in.} \quad (\text{use 6 in.})$$

The connection is shown in Fig. 8-11(b).

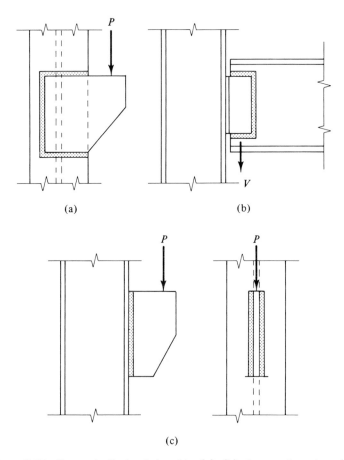

(a)

(b)

(c)

Figure 8-12   Eccentrically loaded welds; (a), (b) shear and torsion; (c) shear and bending.

8-8 ECCENTRICALLY LOADED WELDS

Two types of eccentrically loaded welds will be discussed: shear and torsion and shear and bending, as illustrated in Fig. 8-12. The elastic analysis of the welded shear and torsion connection is similar to the eccentrically loaded bolted connection of Sec. 7-8.

Shear and Torsion: Elastic Analysis

Consider the eccentrically loaded weld shown in Fig. 8-13(a). The downward force has an eccentricity e measured from the centroid of the weld group. The load is replaced by a statically equivalent system—a downward force P acting through the centroid [Fig. 8-13(b)] and a torsion $M = Pe$ about the centroid [Fig. 8-13(c)].

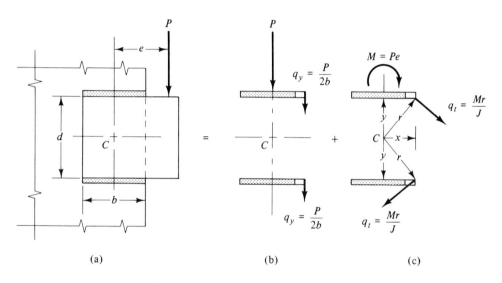

(a)   (b)   (c)

Figure 8-13 Eccentrically loaded weld.

For the load acting through the centroid each unit length of weld is assumed to support an equal part of the load. Therefore, the force on each inch of weld will be a downward force equal to

$$q_y = \frac{P}{2b} \qquad (8\text{-}1)$$

as shown in Fig. 8-13(b).

The torsion is assumed to produce forces that are normal and proportional to the distance from the centroid of the weld group to each unit length of weld as shown in Fig. 8-13(c). The torsion force on each inch of weld is

$$q_t = \frac{Mr}{J} \tag{8-2}$$

and the x and t components of the torsion force are given by

$$q_{tx} = \frac{My}{J} \qquad q_{ty} = \frac{Mx}{J} \tag{8-3}$$

where J is the polar moment of inertia of the weld group. The direction of the torsion forces is determined by inspection.

To compute the polar moment of inertia of the weld group, the weld is assumed to be composed of lines of weld along the edges where the welds were placed as shown in Fig. 8-14. The lines are assumed to have a unit width.

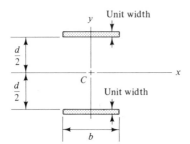

Figure 8-14

The moment of inertia of the two line welds about the x axis from the parallel axis theorem is given by

$$I_x = I_{xc} + Ad_y^2 = 0 + 2\left[b\left(\frac{d}{2}\right)^2 \right] = \frac{bd^2}{2}$$

The moment of inertia about the y axis from the formula for a rectangle of unit width is given by

$$I_y = 2\left(\frac{b^3}{12}\right) = \frac{b^3}{6}$$

Therefore, the polar moment of inertia

$$J = I_x + I_y = \frac{b^3}{6} + \frac{bd^2}{2} = \frac{b(b^2 + 3d^2)}{6} \tag{a}$$

Example 8-5

Determine the maximum force in k/in. on the weld group shown in Fig. 8-15(a) and the weld size required using the SMAW process and E60 electrodes. Assume that the plate does not control weld size and is made of A36 steel.

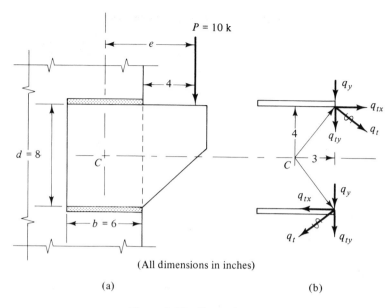

(All dimensions in inches)

(a)   (b)

Figure 8-15   Example 8-5.

Solution.   The centroid of the weld group is located by inspection as shown in the figure. Maximum forces occur on the right-hand ends of both welds. The forces are equal.

The downward force given by Eq. (8-1) and the horizontal and vertical forces given by Eq. (8-3) are illustrated in Fig. 8-15(b).

From Fig. 8-15(a) the eccentricity

$$e = \frac{6}{2} + 4 = 7 \text{ in.}$$

and the external torque

$$M = Pe = 10(7) = 70 \text{ k-in.}$$

From Eq. (a) the polar moment of inertia

$$J = \frac{b(b^2 + 3d^2)}{6} = \frac{6[(6)^2 + 3(8)^2]}{6} = 228 \text{ in.}^3$$

From Eqs. (8-1) and (8-3),

$$q_y = \frac{P}{2b} = \frac{10}{2(6)} = 0.833 \text{ k/in.}$$

$$q_{tx} = \frac{My}{J} = \frac{70(4)}{228} = 1.228 \text{ k/in.}$$

TABLE 8-5 Properties of Weld Groups

Case	Centroid	Section Modulus, $S_x = I_x/\bar{y}$	Polar Moment of Inertia, J, About Centroid
1.		$S = \dfrac{d^2}{6}$	$J = \dfrac{d^3}{12}$
2.		$S = \dfrac{d^2}{3}$	$J = \dfrac{d(3b^2 + d^2)}{6}$
3.		$S = bd$	$J = \dfrac{b(3d^2 + b^2)}{6}$
4.	$\bar{y} = \dfrac{d^2}{2(b + d)}$ $\bar{x} = \dfrac{b^2}{2(b + d)}$	$S = \dfrac{4bd + d^2}{6}$	$J = \dfrac{(b + d)^4 - 6b^2d^2}{12(b + d)}$
5.	$\bar{x} = \dfrac{b^2}{2b + d}$	$S = bd + \dfrac{d^2}{6}$	$J = \dfrac{8b^3 + 6bd^2 + d^3}{12} - \dfrac{b^4}{2b + d}$
6.	$\bar{y} = \dfrac{d^2}{b + 2d}$	$S = \dfrac{2bd + d^2}{3}$	$J = \dfrac{b^3 + 6b^2d + 8d^3}{12} - \dfrac{d^4}{2d + b}$
7.		$S = bd + \dfrac{d^2}{3}$	$J = \dfrac{(b + d)^3}{6}$
8.	$\bar{y} = \dfrac{d^2}{b + 2d}$	$S = \dfrac{2bd + d^2}{3}$	$J = \dfrac{b^3 + 8d^3}{12} - \dfrac{d^4}{b + 2d}$
9.		$S = bd + \dfrac{d^2}{3}$	$J = \dfrac{b^3 + 3b^2 + d^3}{6}$

and

$$q_{ty} = \frac{Mx}{J} = \frac{70(3)}{228} = 0.921 \text{ k/in.}$$

The resultant of the three forces is given by

$$R = \sqrt{q_{tx}^2 + (q_y + q_{ty})^2}$$
$$= \sqrt{(1.228)^2 + (0.833 + 0.921)^2}$$
$$= 2.14 \text{ k/in.}$$

The capacity of a fillet weld using the SMAW process and E60 electrode from Table 8-3: $q_v = 0.80D$. Thus

$$0.80D = 2.14 \qquad D = 2^+ \quad (\text{use } 3)$$

Use a $\frac{3}{16}$-in. weld—or larger.

The formula for the polar moment of inertia is also given as case 3 in Table 8-5, Properties of Weld Groups.

Example 8-6

Determine the maximum force in k/in. on the weld group shown [Fig. 8-16(a)] and the weld size required using the SAW process and F7X flux electrode combination. Assume that the plate does not control the weld size and is made of A36 steel.

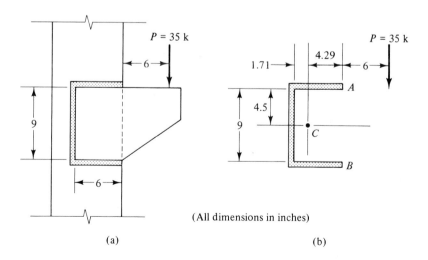

(All dimensions in inches)

(a)   (b)

Figure 8-16   Example 8-6.

Solution. The centroid and polar moment of inertia of the weld group are determined from formulas for case 5, Table 8-5. With $b = 6$ in. and $d = 9$ in.,

$$x = \frac{b^2}{2b + d} = \frac{(6)^2}{2(6) + 9} = 1.71 \text{ in.}$$

and

$$J = \frac{8b^3 + 6bd^2 + d^3}{12} - \frac{b^4}{2b + d}$$

$$= \frac{8(6)^3 + 6(6)(9)^2 + (9)^3}{12} - \frac{(6)^4}{2(6) + 9} = 386 \text{ in.}^3$$

The centroid is located as shown in Fig. 8-16(b). Thus

$$e = 6 + 4.29 = 10.29 \text{ in.}$$

and the external torque

$$M = Pe = 35(10.29) = 360 \text{ k-in.}$$

The maximum force will occur on the right-hand end of the top and bottom welds (points A and B). The resultant forces are equal. From Eqs. (8-1) and (8-3) the force components at A are

$$q_y = \frac{P}{L} = \frac{35}{21} = 1.667 \text{ k/in.}$$

$$q_{tx} = \frac{My}{J} = \frac{360(4.5)}{386} = 4.197 \text{ k/in.}$$

and

$$q_{ty} = \frac{Mx}{J} = \frac{360(4.29)}{386} = 4.001 \text{ k/in.}$$

The resultant force at A is given by

$$q_A = \sqrt{q_{tx}^2 + (q_y + q_{ty})^2}$$

$$= \sqrt{(4.197)^2 + (1.667 + 4.001)^2}$$

$$= 7.05 \text{ k/in.}$$

From Table 8-3 for the SAW process and F7X flux electrode combination $q_v = 1.31D$. Thus

$$1.31D = 7.05 \qquad D = 5^+ \quad (\text{use } 6)$$

Use a $\frac{3}{8}$-in. weld.

Shear and Torsion: Ultimate Strength Analysis

The ultimate strength or plastic method for welds follows the same reasoning used for bolted joints. The tables in the AISC *Manual* are based on the ultimate strength analysis. The use of the tables is illustrated in the following examples.

Example 8-7

Rework Example 8-6 using the ultimate strength analysis (AISC *Manual*).

Solution.   This weld group is given in Table XXIII of the AISC *Manual*, p. 4-80. With $d = l = 9$ in., $e = al = 10.29$ in., and $b = kl = 6$ in.: $a = 1.143$ and $k = 0.667$. The coefficient C is determined by double interpolation. The interpolation is organized in the following table, where the interpolated values are shown in boldface type.

	k		
a	0.6	**0.667**	0.7
1.000	0.711		0.796
1.143	**0.637**	**0.689**	**0.715**
1.200	0.608		0.683

The coefficient $C = 0.698$ is for the SMAW process with an E70 electrode. For the SAW process and F7X flux electrode combination the coefficient C_1 is given by

$$C_1 = \frac{q_v \text{ (SAW process—F7X flux electrode)}}{q_v \text{ (SMAW process—E7O electrode)}}$$

From Table 8-3,

$$C_1 = \frac{1.31D}{0.93D} = 1.41$$

Therefore, the size of the weld

$$D = \frac{P}{CC_1 l} = \frac{35}{0.689(1.41)(9)} = 4.0 \quad \text{(use 4)}$$

Use a $\frac{1}{4}$-in. weld.

In Example 8-7 the elastic analysis gives a result approximately 35 percent larger than the ultimate strength analysis.

Example 8-8

With the ultimate strength method (AISC *Manual*) determine the weld size required for the eccentrically loaded connection as shown in Fig. 8-17.

(a) Use the SMAW process and E8O electrodes.

(b) Use the SAW process and F8X flux electrode combination.

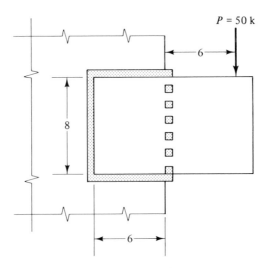

(All dimensions in inches)

Figure 8-17   Example 8-8.

Solution.   This weld group is given in Table XXI of the AISC *Manual*, p. 4-78. With $d = l = 8$ in., $e = al = 9$ in., and $b = kl = 6$ in.: $a = 1.125$ and $k = 0.75$. The coefficient C is determined by double interpolation as shown in Example 8-7. $C = 1.12$.

(a) From the AISC *Manual*, p. 4-74, for the SMAW process and E8O electrode: $C_1 = 1.14$. Therefore, the size of weld

$$D = \frac{P}{CC_1 l} = \frac{50}{1.12(1.14)(8)} = 4^+ \quad \text{(use 5)}$$

Use a $\frac{5}{16}$-in. weld.

(b) For the SAW process and F8X electrode combination

$$C_1 = \frac{q_v \text{ (SAW process—F8X flux electrode)}}{q_v \text{ (SMAW process—E7O electrode)}}$$

$$= \frac{1.50D}{0.93D} = 1.61$$

Therefore, the size of the weld

$$D = \frac{P}{CC_1l} = \frac{50}{1.12(1.61)(8)} = 3^+ \quad (\text{use } 4)$$

Use a $\frac{1}{4}$-in. weld.

Shear and Bending: Elastic Method

The combined shear and bending force per unit length of weld is calculated by adding vectorially the nominal shear force and bending force per unit length of weld.

Consider the welded bracket illustrated in Fig. 8-18. The nominal shear force per inch of weld is given by

$$q_y = \frac{P}{2d} \tag{8-4}$$

and the maximum bending force at the top and bottom (points A and B) of the weld group

$$q_b = \frac{M}{S_z} \tag{8-5}$$

The section modulus of the two line welds about the z axis is given by

$$S_z = \frac{I_z}{c} = \frac{2d^3/12}{d/2} = \frac{d^2}{3} \tag{a}$$

Since the shear force and bending force are at right angles with each other, the resultant at point A is given by

$$q_A = \sqrt{q_y^2 + q_b^2}$$

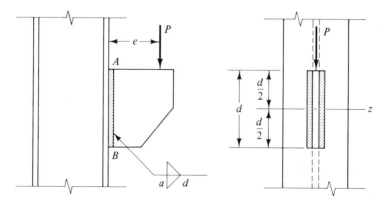

Figure 8-18   Example 8-9.

Example 8-9

If $P = 15$ k, $e = 6$ in., and $d = 8$ in. determine the weld size required for the connection shown in Fig. 8-18 if the SMAW process is used with E70 electrodes.

(a) Use the elastic method.

(b) Use the ultimate strength method (AISC *Manual*).

Solution. (a) The section modulus from Eq. (a)

$$S_z = \frac{d^2}{3} = \frac{(8)^2}{3} = 21.3 \text{ in.}^2$$

and the moment $M = Pe = 15(6) = 90$ k-in. From Eqs. (8-4) and (8-5),

$$q_y = \frac{P}{2d} = \frac{15}{2(8)} = 0.938 \text{ k/in. of weld}$$

and

$$q_b = \frac{M}{S_z} = \frac{90}{21.3} = 4.23 \text{ k/in. of weld}$$

The resultant force at point A

$$q_A = \sqrt{q_y^2 + q_b^2} = \sqrt{(4.23)^2 + (0.938)^2}$$
$$= 4.33 \text{ k/in. of weld}$$

From Table 8-3 for the SMAW process with E70 electrodes: $q = 0.93D$. Thus

$$0.93D = 4.33 \qquad D = 4^+ \quad (\text{use 5})$$

Use a $\frac{5}{16}$-in. weld.

The formula for the section modulus is given as case 2 in Table 8-5, Properties of Weld Groups.

(b) Ultimate strength method (AISC *Manual*). This weld group is given in Table XIX of the AISC *Manual*, p. 4-76. With $l = 8$ in. and $al = 6$ in., $a = 0.75$ and $k = 0$. The coefficient by interpolation is $C = 0.551$. For the SMAW process and E70 electrode from AISC *Manual*, p. 4-74, $C_1 = 1.0$. Therefore, the weld size

$$D = \frac{P}{CC_1 l} = \frac{15}{0.551(1.0)(8)} = 3^+ \quad (\text{use 4})$$

Use a $\frac{1}{4}$-in. weld.

8-9 SHEAR CONNECTIONS FOR BUILDING FRAMES

The shear or web-farming connection is used to connect beams to girders or columns and provide simple beam supports. The web-framing angles are

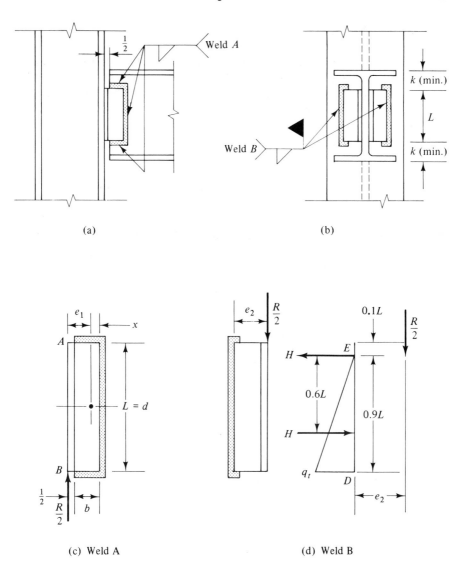

(a)

(b)

(c) Weld A

(d) Weld B

Figure 8-19   Welded web-framing beam connection.

usually shop welded to the beam [Fig. 8-19(a)] and field welded to the column or girder [Fig. 8-19(b)]. Erection bolts may be used while the connection is being field welded. Another method is to shop weld the angles to the beam and field bolt them to the column or girder. See the AISC *Manual,* Table III, p. 4-28, for combined bolted and welded connections. All welded connections are discussed in the following paragraphs.

Design Process

Welded design can be separated into two parts, the shop weld and the field weld. The shop weld, weld A in Fig. 8-19(c), is an eccentrically loaded weld subject to shear and torsion. Welds of this type were discussed in Sec. 8-8.

The field weld, weld B in Fig. 8-19(d), is also eccentrically loaded. However, the framing angles are not free to rotate about the centroid of the weld group. The tops of the angles bear against the beam web, so each other and the bottom of the angles rotate away from the beam web.

For design purposes, we assume that the torsion causes the angles to bear against themselves for a distance $0.1L$ from the top and to produce a triangular distribution of forces in the weld over the remaining distance of $0.9L$. The resultant horizontal bearing force is assumed to act at point E and the resultant of the distributed horizontal torsional forces acts at the third point of the triangular distribution of forces as shown. The resultant of the distributed torsional forces is given by

$$H = \frac{q_t 0.9L}{2} \tag{a}$$

From moment equilibrium about point E,

$$M_E = \frac{R}{2} e_2 - H0.6L = 0 \tag{b}$$

Solving for q_t from Eqs. (a) and (b) yields

$$q_t = \frac{Re_2}{0.54L^2} \tag{8-6}$$

where q_t is the maximum horizontal torsion force per unit length of weld. The maximum force occurs at D.

The shear force per unit length of weld

$$q_y = \frac{R}{2L} \tag{8-7}$$

The shear force and horizontal torsion force are at right angles to each other. Therefore, the resultant at point D

$$q_D = \sqrt{q_y^2 + q_t^2}$$

Example 8-10

Select the welds and framing angles to connect a W27 × 94 beam to a W12 × 152 column. The rolled sections are of A36 steel and the reactive force to be transmitted is $R = 90$ k. Use the SMAW process and E70 electrodes.

Solution

    W27 × 94 beam: $t_w = 0.490$ in., $T = 24$ in.

    W12 × 152 column: $t_f = 1.400$ in.

Use two L3 × 3 × t × L framing angles. To estimate the length of angle required, we calculate the length of $\frac{1}{4}$-in. weld required to support the reaction. With $D = 4$, from Table 8-3,

$$q = 0.93D = 0.93(4) = 3.72 \text{ k/in.}$$

The length of weld required

$$L = \frac{R}{2q} = \frac{90}{2(3.72)} = 12.1 \text{ in.} < T = 24 \text{ in.} \qquad \text{OK}$$

Use $L = d = 12$ in.

    Weld A. Referring to Fig. 8-19(c), we have

$$b = \text{leg length} - \text{setback} = 3.0 - 0.5 = 2.5 \text{ in.}$$

From case 5, Table 8-5,

$$x = \frac{b^2}{(2b + d)} = \frac{(2.5)^2}{2(2.5) + 12} = 0.368 \text{ in.}$$

$$J = \frac{8b^3 + 6bd^2 + d^3}{12} - \frac{b^4}{2b + d}$$

$$= \frac{8(2.5)^3 + 6(2.5)(12)^2 + (12)^3}{12} - \frac{(2.5)^4}{2(2.5) + 12}$$

$$= 332 \text{ in.}^2$$

$$e_1 = \text{leg length} - x = 3.0 - 0.368 = 2.632 \text{ in.}$$

The maximum force occurs at point A. From Eqs. (8-1) and (8-3),

$$q_y = \frac{R}{L} = \frac{45}{12 + 2(2.5)} = 2.647 \text{ k/in.}$$

$$q_{tx} = \frac{My}{J} = \frac{45(2.632)(6)}{332} = 2.140 \text{ k/in.}$$

and

$$q_{ty} = \frac{Mx}{J} = \frac{45(2.632)(2.132)}{332} = 0.761 \text{ k/in.}$$

The resultant at point A

$$q_A = \sqrt{q_{tx}^2 + (q_y + q_{ty})^2}$$

$$= \sqrt{(2.140)^2 + (2.647 + 0.761)^2}$$

$$= 4.02 \text{ k/in.}$$

From Table 8-3, $q_v = 0.93D$. Thus

$$0.93D = 4.02 \qquad D = 4^+ \quad (\text{use } \tfrac{5}{16}\text{-in. weld})$$

Check of web shear capacity

$$t_w 0.4F_y = 0.49(0.4)(36)$$
$$= 7.06 \text{ k/in.} > 0.93(5) = 4.65 \text{ k/in.} \quad \text{OK}$$

Weld B. Referring to Fig. 8-19(d), we find that the eccentricity $e_2 = 3.0$ in. From Eqs. (8-6) and (8-7),

$$q_{th} = \frac{Re_2}{0.54L^2} = \frac{90(3.0)}{0.54(12)^2} = 3.47 \text{ k/in.}$$

and

$$q_y = \frac{R}{2L} = \frac{90}{2(12)} = 3.75 \text{ k/in.}$$

The resultant at point D

$$q_D = \sqrt{q_{th}^2 + q_y^2} = \sqrt{(3.47)^2 + (3.75)^2} = 5.11 \text{ k/in.}$$

From Table 8-3, $q = 0.93D$. Therefore,

$$0.93D = 5.11 \qquad D = 5^+ \quad (\text{use } \tfrac{3}{8}\text{-in. weld})$$

The minimum fillet weld size that can be used with a flange thickness $t_f = 1.4$ in. from Table 8-4 is $\tfrac{5}{16}$ in. $< a = \tfrac{3}{8}$ in.

The minimum angle thickness

$$t = \text{weld size} + \frac{1}{16} = \frac{3}{8} + \frac{1}{16} = \frac{7}{16} \text{ in.}$$

Use an angle thickness of $\tfrac{1}{2}$ in.

Check of angle shear capacity

$$t_L 0.4F_y = 0.5(0.4)(36)$$
$$= 7.2 \text{ k/in} > 0.93D = 0.93(6) = 5.58 \text{ k/in.} \quad \text{OK}$$

Detail data. Two L3 × 3 × $\tfrac{1}{2}$ × 1 ft-0 in., $F_y = 36$ ksi.

$$\text{Weld A: } a = \frac{5}{16}\text{-in., SMAW process, E70 electrode}$$

$$\text{Weld B: } a = \frac{3}{8}\text{-in., SMAW process, E70 electrode}$$

Example 8-11

Select the welds and framing angles to connect a W18 × 55 beam to a W33 × 141 girder. The sections are A36 steel and the reaction on the beam is $R = 85$ k. Use the SMAW process, E70 electrodes, and the AISC *Manual* tables.

Solution

W18 × 55 beam: $t_w = 0.390$ in., $T = 15.5$ in.

W33 × 141 girder: $t_w = 0.605$ in.

Enter Table IV, AISC *Manual*, p. 4-36, and select for:

Weld A: $\frac{1}{4}$-in. weld has a capacity of 112 k for a beam web thickness of 0.51 in. The capacity for a web thickness of 0.390 in. is 85.6 k > 85 k.   (OK)

Weld B: $\frac{5}{16}$-in. weld has a capacity of 103 k > 85 k.   (OK)

Use two L3 × 3 × $\frac{3}{8}$ × 1 ft-2 in.

Check of web shear capacity

$$t_w 0.4 F_y = 0.390(0.4)(36)$$

$$= 5.62 \text{ k/in.} > 0.93D = 0.93(4) = 3.72 \text{ k/in.} \quad \text{OK}$$

The minimum size for weld B with a girder web thickness of $t_w = 0.605$ in. from Table 8-4 is $\frac{1}{4}$ in. < weld size $a = \frac{5}{16}$ in.   (OK)   The minimum angle thickness

$$t = \text{weld size} + \frac{1}{16} = \frac{5}{16} + \frac{1}{16} = \frac{3}{8} \text{ in.}$$

Check of angle shear capacity

$$t_L 0.4 F_y = 0.375(0.4)(36)$$

$$= 5.4 \text{ k/in.} > 0.93D = 0.93(5) = 4.65 \text{ k/in.} \quad \text{OK}$$

Detail data. Two L3 × 3 × $\frac{3}{8}$ × 1 ft-2 in., $F_y = 36$ ksi.

Weld A: $a = \dfrac{1}{4}$-in. SMAW process, E70 electrode

Weld B: $a = \dfrac{5}{16}$-in. SMAW process, E70 electrode

8-10 SEATED BEAM CONNECTIONS

The design of a welded seated beam connection follows the same procedure as the bolted seated beam connection discussed in Sec. 7-12. In place of bolts that were used to attach the seat to a column flange or web, welds are placed along the edges of the seat as shown in Fig. 8-20. The welds are acted on by shear and bending and are designed by the methods of Sec. 8-8.

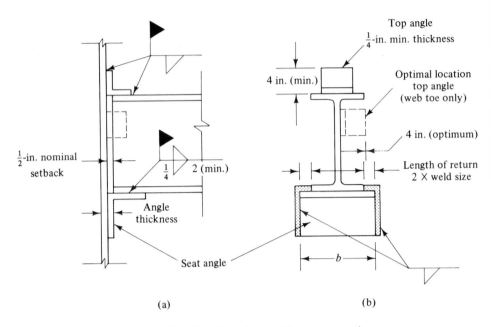

Figure 8-20 Unstiffened seated beam connection.

Example 8-12

Design the seated connection to support a W18 × 40 beam of A36 steel. The reactive force is $R = 26$ k. The beam seat is connected to a column web by welds. Determine the weld size required if the SMAW process is used with E60 electrodes.

Solution. A similar seated beam connection with bolts was designed in Example 7-12. In this example one of the legs of the angle will be lengthened from 4 in. to 6 in. to provide longer edges for the weld.

The seat angle, critical section, eccentricity, reactive force, and weld orientation are illustrated in Fig. 8-21. The welds are acted on by shear and bending. From Eqs. (8-4) and (8-5) the nominal shear force per inch of weld for the weld group shown in Fig. 8-20(b) is

$$q_y = \frac{R}{2d} = \frac{26}{2(6)} = 2.17 \text{ k/in.}$$

and the maximum bending force at the top, point A, is

$$q_b = \frac{M}{S_z} = \frac{M}{d^2/3} = \frac{26(0.466)(3)}{(6)^2} = 1.010 \text{ k/in.}$$

where the section modulus is from Table 8-5. The resultant force at point A

$$q_A = \sqrt{q_y^2 + q_b^2} = \sqrt{(2.17)^2 + (1.010)^2} = 2.39 \text{ k/in.}$$

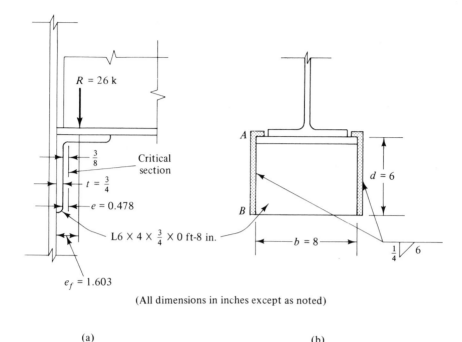

(All dimensions in inches except as noted)

(a)   (b)

Figure 8-21   Example 8-12.

For the SMAW process with an E60 electrode from Table 8-3, $q = 0.80D$. Thus

$$0.80D = 2.39 \qquad D = 2^{+} \quad (\text{use } \tfrac{1}{4} \text{ in.})$$

The minimum weld size for the $\tfrac{3}{4}$-in. angle is $\tfrac{1}{4}$ in. Use seat angle $6 \times 4 \times \tfrac{3}{4} \times$ 0 ft-8 in. with the outstanding leg = 4 in. and two $\tfrac{1}{4}$-in. welds 6 in. long made with E60 electrodes by the SMAW process. Provide $\tfrac{1}{2}$-in. returns at the top of the welds.

Example 8-13

Design the seated connection to support a W21 × 50 beam of A36 steel. The reactive force is $R = 26$ k. The beam seat is connected to a column flange by welds. Determine the beam seat angle and weld size required if the SMAW process is used with E70 electrodes.

Solution.   W21 × 50 beam: $t_w = \tfrac{3}{8}$ in.

Table VI-A: With a length $b = 8$ in. and a web thickness $t_w = \tfrac{3}{8}$ in., select a $\tfrac{5}{8}$-in.-thick angle that has a capacity of 26.3 k > 26 k. (OK)

Table VI-C: Satisfactory weld capacities appear under 5- through 8-in. leg angles, all of which are available in $\tfrac{5}{8}$-in. thickness. Angle 7 × 4 (capacity = 28.5 k, $\tfrac{1}{4}$-in. weld) and 6 × 4 (capacity = 27.3 k, $\tfrac{5}{16}$-in. weld) are equally suitable.

Angle 6×4 is chosen because of material savings and $\frac{5}{16}$-in. weld can still be made in a single pass.

Detail data: One $L6 \times 4 \times \frac{5}{8} \times 0$ ft-8 in. with $\frac{5}{16}$-in. welds (SMAW process, E70 electrodes). Top or side angle as required.

8-11 STIFFENED SEATED BEAM CONNECTIONS

Except for the stiffening used, the discussion for bolted stiffened seats in Sec. 7-13 also applies to welded seats. The stiffened seat is either a structural tee section or two plates welded together to form a tee shape. The design of the welded stiffened beam seat is illustrated in the following example.

Example 8-14

Design a stiffened beam seat connection for a W21 × 94 beam of A36 steel. The reactive force is $R = 70$ k. The beam seat is connected to a column flange by welds. Use a structural tee section as shown in Fig. 8-22. Determine the weld size required if the SMAW process is used with E70 electrodes.

Solution. Properties of W27 × 94: $t_w = 0.490$ in., $b_f = 9.99$ in., $k = 1\frac{7}{16}$ in. Solving for the required bearing length N from Eq. (5-9) [AISC Formula (1.10-9)] yields

$$N = \frac{R}{0.75 F_y t_w} - k = \frac{70}{0.75(36)(0.490)} - 1\frac{7}{16}$$

$$= 3.85 \text{ in.}$$

Assume a nominal setback of $\frac{1}{2}$ in. plus $\frac{1}{4}$-in. allowance for beam length underrun. Therefore, the erection clearance is 0.75 in.

$$W = \text{erection clearance} + N$$

$$= 0.75 + 3.85 = 4.6 \text{ in.} \quad (\text{use 5 in.})$$

The eccentricity is assumed to be the erection clearance plus half the *remaining* seat length. Thus

$$e = 0.75 + \frac{5 - 0.75}{2} = 2.875 \text{ in.}$$

The moment on the connection

$$M = Pe = 70(2.875) = 201 \text{ k-in.}$$

The length of weld can be estimated by dividing the reaction by the strength of a $\frac{3}{16}$-in. fillet weld (E70 electrode). From Table 8-3 for $D = 3$, $q = 0.93D = 0.93(3) = 2.79$ k/in. The length of weld required

$$L = \frac{R}{q} = \frac{70}{2.79} = 25 \text{ in.}$$

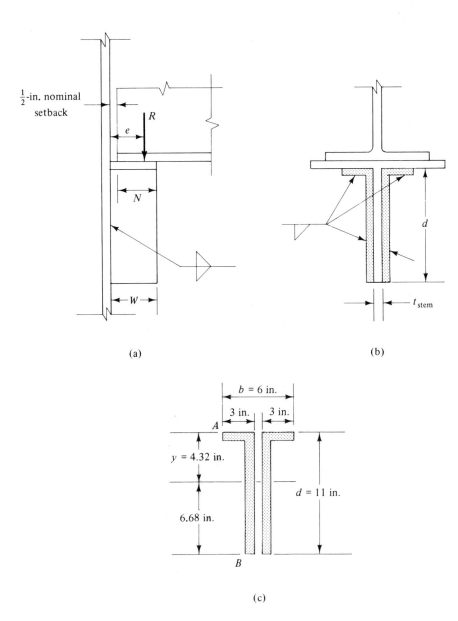

(a)   (b)

(c)

Figure 8-22   Example 8-14.

Assume welds as shown in Fig. 8-22(c). For case 8 from Table 8-5 with $d = 11$ in. and $b = 6$ in.,

$$y = \frac{d^2}{b + 2d} = \frac{(11)^2}{6 + 2(11)} = 4.32 \text{ in.}$$

$$S_z = \frac{2bd + d^2}{3} = \frac{2(6)(11) + (11)^2}{3} = 84.3 \text{ in.}^2$$

From Eq. (8-4) the nominal shear force is given by

$$q_y = \frac{R}{b + 2d} = \frac{70}{6 + 2(11)} = 2.50 \text{ k/in.}$$

and from Eq. (8-5) the maximum bending force at point A of the weld group is

$$q_b = \frac{M}{S_z} = \frac{201}{84.3} = 2.39 \text{ k/in.}$$

The resultant force at point A

$$q_A = \sqrt{q_y^2 + q_b^2} = \sqrt{(2.50)^2 + (2.39)^2} = 3.46 \text{ k/in.}$$

From Table 8-3 for the SMAW process with E70 electrodes $q = 0.93D$. Thus

$$0.93D = 3.46 \qquad D = 3^+ \quad (\text{use } \tfrac{1}{4}\text{-in. weld})$$

Requirements for structural tee section. The structural tee stem thickness t_w should satisfy the following:

$$t_s \geq t_w(\text{beam web}) = 0.490 \text{ in.}$$

$$t_s \geq 2(\text{weld size}) = 2\left(\frac{1}{4}\right) = 0.5 \text{ in.}$$

and according to AISC Specification 1.9.1.2 to prevent buckling

$$t_s = \frac{w\sqrt{F_y}}{127} = \frac{5(6)}{127} = 0.236 \text{ in.}$$

Also, the structural tee flange should be at least 1 in. wider than the beam flange

$$b_f = b_f(\text{beam}) + 1.0 = 9.99 + 1.0 = 10.99 \text{ in.}$$

and the depth 1 in. larger than d of the weld

$$d = d(\text{weld}) + 1.0 = 11.0 + 1.0 = 12 \text{ in.}$$

Use a WT12 × 52: $t_w = 0.55$ in., $b_f = 12.750$ in., $d = 12.030$ in. and $\tfrac{1}{4}$-in. fillet welds. A larger weld size would be required if the column flange thickness exceeds $\tfrac{3}{4}$ in. (Table 8-4).

The design of stiffened seats composed of two plates welded together to form a tee shape is discussed in the AISC *Manual.*

8-12 MOMENT-RESISTING CONNECTIONS

Moment-resisting connections were discussed in Sec. 7-23 in connection with bolts.

Flange and Vertical Web Plate Moment Connection

The moment connection with flange and vertical web plate can be used as either a rigid or semirigid connection. In the connection illustrated in Fig. 8-23, the shear force V is transmitted to a column by a vertical plate that is shop welded to the column flange and field bolted to the beam web. The

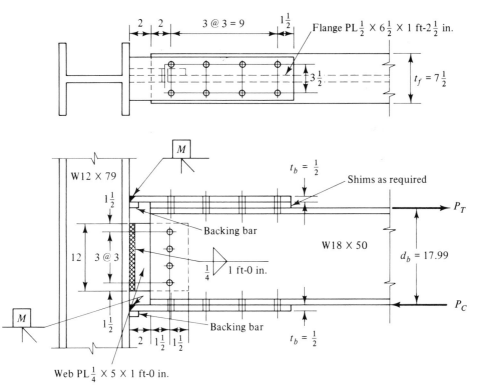

(All dimensions in inches except as noted)

Figure 8-23   Example 8-15. Moment connection with flange and vertical web plate.

bending moment is transmitted to the column flange by plates that are shop welded to the column flange and field bolted to the beam flanges. In some cases the flange plates may be field welded to the beam flanges.

The design procedure is illustrated in the following example.

Example 8-15

Design a moment connection with flange and vertical web plates to join a W18 × 50 beam to a W12 × 79 column. The design moment $M = 95$ k-ft and the shear $V = 37$ k are the result of dead and live load only. The members are made of A36 steel with $F_t = 22$ ksi (for beam $F_b = 24$ ksi). Use $\frac{3}{4}$-in. A325-N bolts and welds made by the SMAW process with E70 electrodes.

Solution

W18 × 50 beam: $d = 17.99$ in., $b_f = 7.495$ in., $t_f = 0.570$ in., $t_w = 0.355$ in.

W12 × 79 column: $d = 12.38$ in., $b_f = 12.080$ in., $t_f = 0.735$ in., $t_w = 0.470$ in., $k = 1\frac{7}{16}$ in.

Beam flange area reduction

$$A_g = 7.495(0.570) = 4.27 \text{ in.}^2$$

Assume two rows of $\frac{3}{4}$-in.-diameter connecting bolts

$$A_n = 4.27 - 2\left(\frac{3}{4} + \frac{1}{8}\right)(0.570) = 3.27 \text{ in.}^2$$

The percent reduction

$$\frac{(4.27 - 3.27)(100)}{4.27} = 23.4\% > 15\% \ (8.4\% \text{ excess})$$

From AISC Specification 1.10.1 the moment of inertia of the flanges must be reduced by 8.4 percent excess (Sec. 5-8). The net moment of inertia

$$I_{net} = 800 - 2(0.084)(4.27)\left(\frac{17.99 - 0.570}{2}\right)^2$$

$$= 745.6 \text{ in.}^4$$

$$S_{net} = \frac{I_{net}}{c} = \frac{745.6}{17.99/2} = 82.9 \text{ in.}^3$$

The bending stress

$$f_b = \frac{M}{S_{net}} = \frac{95(12)}{82.9} = 13.75 \text{ ksi} < F_b = 24 \text{ ksi} \qquad \text{OK}$$

Horizontal force on beam flanges

$$P = P_T = P_C = \frac{M}{d} = \frac{95(12)}{17.99} = 63.4 \text{ k}$$

Design of flange plates. The required area of the flange plates

$$A_n = \frac{P}{0.5F_u} = \frac{63.4}{0.5(58)} = 2.19 \text{ in.}^2$$

and

$$A_g = \frac{P}{0.6F_y} = \frac{63.4}{22} = 2.88 \text{ in.}^2$$

Try a $\frac{1}{2}$-in. plate. The width of the plate based on the net area

$$b = \frac{2.19}{0.5} + 2\left(\frac{3}{4} + \frac{1}{8}\right) = 6.13 \text{ in.}$$

and based on the gross area

$$b = \frac{2.88}{0.5} = 5.76 \text{ in.}$$

Use $\frac{1}{2} \times 6\frac{1}{2}$ flange plates on the top and bottom.

Flange connection. The shear capacity of a $\frac{3}{4}$-in. A325-N bolt in single shear from AISC *Manual*, Table I-D, is $r_v = 9.3$ k. The bearing from Table I-E for a $\frac{1}{2}$-in. plate with 3-in. bolt spacing is 32.6 k. Therefore, the number of bolts required

$$N = \frac{P}{r_v} = \frac{63.4}{9.3} = 6^+$$

Use eight $\frac{3}{4}$-in.-diameter A325-N bolts.

Web connection. The shear capacity of a $\frac{3}{4}$-in. bolt in single shear is 9.3 k. The bearing from Table I-E for a web thickness $t_w = 0.355$ in. with 3-in. spacing is $65.3(0.355) = 23.2$ k > 9.3 k. The number of bolts required

$$N = \frac{V}{r_v} = \frac{37}{9.3} = 3^+$$

Use four $\frac{3}{4}$-in.-diameter A325-N bolts.

Design of web plate. Try a 12-in. web plate with 3-in. bolt spacing and edge distance $l_v = 1\frac{1}{2}$ in.

$$l_{\text{net}} = 12 - 4\left(\frac{3}{4} - \frac{1}{8}\right) = 8.5 \text{ in.}$$

The allowable stress from AISC Specification 1.5.1.2.2 is

$$F_v = 0.3F_u = 0.3(58) = 17.4 \text{ ksi}$$

The plate thickness

$$t = \frac{R}{l_{\text{net}}F_v} = \frac{37}{8.5(17.4)} = 0.25 \text{ in.}$$

Try a $\frac{1}{4}$-in. plate.

For $F_u = 58$ ksi, $l_v = 1\frac{1}{2}$ in. for AISC *Manual*, Table I-E (edge distance controls), the bolt capacity $= 10.9$ k $> r_v = 9.3$ k.   (OK)

For edge distances for the plate $l_v = l_h = 1\frac{1}{2}$ in. from AISC *Manual*, Table I-F, the bearing capacity for the $\frac{1}{4}$-in. plate $4(\frac{1}{4})(43.5) = 43.5$ k > 37 k.   (OK)   Use a web plate $\frac{1}{4} \times 5 \times 12$ welded to the column flange with $\frac{1}{4}$-in. fillet welds full length both sides.

Check of web shear in column. From AISC Specification Commentary 1.5.1.2,

$$A_{bc} = d_b d_c = [17.99 + 2(0.5)](12.38) = 235.2 \text{ in.}^2$$

$$t_{min} = \frac{32M}{A_{bc}F_y} = \frac{32(95)}{235.1(36)} = 0.359 \text{ in.} < 0.470 \text{ in.}   \text{OK}$$

The column web need not be reinforced.

Check of web stiffeners for columns. From AISC Specification 1.15.5.2,

$$P_{bf} = \frac{5}{3} P = \frac{5}{3} (63.4) = 105.7 \text{ k}$$

when the computed forces are due to live and dead load only. (For live and dead loads in conjunction with wind or earthquake forces, $P_{bf} = 4P/3$.)

From AISC Formula (1.15-1),

$$A_{st} = \frac{P_{bf} - F_{yc}t(t_b + 5k)}{F_{yst}}$$

$$= \frac{105.7 - 36(0.470)[0.5 + 5(1.4375)]}{36}$$

$$= -0.677 \text{ in.}^2$$

Since A_{st} is negative a further check is required. From AISC Formula (1.15-2) for the compression flange connection,

$$d_c = d - 2k = 12.38 - 2(1.4375) = 9.505 \text{ in.}$$

$$d_c = \frac{4100t^3\sqrt{F_{yc}}}{P_{bf}} = \frac{4100(0.5)^3\sqrt{36}}{105.7} = 29.1 \text{ in.}$$

$$9.505 \text{ in.} < 29.1 \text{ in.}   \text{OK}$$

No column web stiffeners are required at the compression flange plate. From AISC Formula (1.15-3) for the tension flange connection,

$$t_f = 0.4\sqrt{\frac{P_{bf}}{F_{yc}}} = 0.4\sqrt{\frac{105.7}{36}} = 0.685 \text{ in.}$$

$$0.735 \text{ in.} > 0.685 \text{ in.}   \text{OK}$$

No column web stiffeners are required at the tension flange plate. The design is shown in Fig. 8-23.

See Example 8-16 and AISC *Manual*, pp. 4-98 through 4-110, for the design of column web stiffeners.

End Plate Moment Connection

The moment connection with a plate shop welded to the end of the beam and field bolted to the flange or web of the column illustrated in Fig. 8-24 is a complex, highly indeterminate connection. The connection is of considerable practical importance. The following design procedure follows the method outlined in the AISC *Manual*.

Example 8-16

Design an end plate moment connection to resist the maximum negative moment and end reaction for a W14 × 48 beam connected to a W14 × 109 column. The members are made of A36 steel. Use A325-N bolts and welds made by the SMAW process with E70 electrodes.

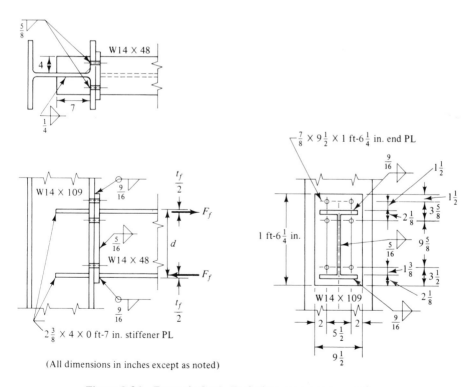

(All dimensions in inches except as noted)

Figure 8-24   Example 8-16. End plate moment connection.

Solution

W14 × 48 beam: $d = 13.75$ in., $t_w = 0.340$ in., $b_f = 8.030$ in., $t_f = 0.595$ in., $T = 11$ in., $k = 1\frac{3}{8}$ in., $S_x = 70.3$ in.^{3}

W14 × 109 column: $d = 14.32$ in., $t_w = 0.525$ in., $b_f = 14.605$ in., $t_f = 0.860$ in., $T = 11\frac{1}{4}$ in., $k = 1\frac{9}{16}$ in.

Maximum negative moment and end reaction. The maximum negative moment

$$M = \frac{F_b S_x}{12} = \frac{24.0(70.3)}{12} = 140.6 \text{ k-ft}$$

The maximum end reaction or shear force

$$V = F_v A_w = 14.4(13.75)(0.340) = 67.3 \text{ k}$$

Bolt design. The flange force due to the moment

$$F_f = \frac{12M}{d - t_f} = \frac{12(140.6)}{13.75 - 0.595} = 128.3 \text{ k}$$

The tensile force is resisted by four bolts at the tension flange. The allowable tensile stress from AISC *Manual,* Table I-A, for A325 bolts is $F_t = 44.0$ ksi. The area required for each bolt to resist the tension

$$A_b = \frac{F_f}{nF_t} = \frac{128.3}{4(44)} = 0.729 \text{ in.}^2$$

The minimum required bolt diameter is $d = 1$ in. with an area $A_b = 0.785$ in.^{2}.

The minimum shear capacity for a 1-in.-diameter A325-N bolt from Table I-D, AISC *Manual,* is $r_v = 16.5$ k. The number of bolts required to support the end shear is

$$N = \frac{V}{r_v} = \frac{67.3}{16.5} = 4^+ \quad (\text{use } 6)$$

Use six 1-in.-diameter A325-N bolts, four at the tension flange and two at the compression flange.

End plate design (see AISC *Manual,* pp. 4-111, 4-112, for equations and definition of symbols). The maximum design plate width

$$b_s = 1.15 b_f = 1.15(8.030) = 9.23 \text{ in.}$$

Use a $9\frac{1}{2}$-in. plate.

The distance from the bolt centerline to the nearest edge of the beam tension flange

$$P_f = d_b + 0.5 = 1.0 + 0.5 = 1.5 \text{ in.}$$

Assume that the fillet weld size $a = \frac{9}{16}$ in. for top flange-to-end plate weld. The effective span

$$P_e = P_f - 0.25d_b - 0.707a$$

$$= 1.5 - 0.25(1.0) - 0.707\left(\frac{9}{16}\right) = 0.852 \text{ in.}$$

$$C_a = 1.13 \qquad \text{(AISC } \textit{Manual}, \text{ Table A, p. 4-113)}$$

$$C_b = \left(\frac{b_f}{b_s}\right)^{1/2} = \left(\frac{8.03}{9.5}\right)^{1/2} = 0.919$$

$$\frac{A_f}{A_w} = 1.115 \qquad \text{(AISC } \textit{Manual}, \text{ Table B, p. 4-113)}$$

$$\alpha_m = C_a C_b \left(\frac{A_f}{A_w}\right)^{1/3} \left(\frac{P_e}{d_b}\right)^{1/4}$$

$$= 1.13(0.919)(1.115)^{1/3}\left(\frac{0.852}{1.0}\right)^{1/4} = 1.035$$

The end plate is designed to resist the moment

$$M_e = \frac{\alpha_m F_f P_e}{4} = \frac{1.035(128.3)(0.852)}{4} = 28.3 \text{ k-in.}$$

The required plate thickness for an allowable bending stress $F_b = 0.75F_y = 27$ ksi,

$$t_s = \left(\frac{6M_e}{b_s F_b}\right)^{1/2} = \left[\frac{6(28.3)}{9.5(27)}\right]^{1/2} = 0.814 \text{ in.}$$

Try a $\frac{7}{8}$-in. $\times$ $9\frac{1}{2}$-in. plate.

Check of plate shear

$$f_v = \frac{F_f}{2b_s t_s} = \frac{128.3}{2(9.5)(0.875)} = 7.71 \text{ ksi} < 14.4 \text{ ksi} \qquad \text{OK}$$

Top flange-to-end plate weld. The required length $L = 2(b_f + t_f) - t_w = 2(8.03 + 0.595) - 0.340 = 16.91$ in. For the SMAW process with E70 electrodes

$$D = \frac{F_f}{0.93L} = \frac{128.3}{0.93(16.91)} = 8^+$$

Use a $\frac{9}{16}$-in. fillet weld.

Beam-to-end plate weld. The minimum-size fillet weld is $\frac{5}{16}$ in. from Table 8-4 (AISC Specification 1.17.2). The minimum length of weld based on the shear capacity of the beam web

$$L = \frac{V}{t_w F_v} = \frac{67.3}{0.340(14.4)} = 13.75 \text{ in.}$$

The required weld to develop maximum bending stress ($F_b = 0.66F_y = 24$ ksi) in the web near the flange

$$D = \frac{t_w F_b}{2(0.93)} = \frac{0.340(24)}{2(0.93)} = 4^+ \quad (\text{use } 5) \qquad \text{OK}$$

Use $\frac{5}{16}$-in. fillet welds continuous on both sides of the beam web.

Column web stiffeners and reinforcement. The horizontal force at possible column stiffeners $F_f = 128.3$ k. From AISC Specification 1.15.5.2

$$P_{bf} = \frac{5}{3} - F_f = \frac{5}{3}(128.3) = 213.8 \text{ k}$$

From AISC Formula (1.15-1),

$$A_{st} = \frac{P_{bf} - F_{yc}t(t_b + 5k)}{F_{yst}}$$

$$= \frac{213.6 - 36(0.525)[0.595 + 5(1.5625)]}{36}$$

$$= 1.52 \text{ in.}^2$$

Stiffeners are required.

From AISC Specification 1.15.5.4, parts (1), (2), and (3), the stiffener width, thickness, and length are given by

$$w_{st} = \frac{b_{bf}}{3} - \frac{t_{cw}}{2} = \frac{8.030}{3} - \frac{0.525}{2}$$

$$= 2.41 \text{ in.} \quad (\text{try } 4 \text{ in.})$$

$$t_{st} = \frac{t_{bf}}{2} = \frac{0.592}{2} = 0.298 \text{ in.} \quad (\text{try } \tfrac{3}{8} \text{ in.})$$

$$l_{st} = \frac{d_c - t_{cf}}{2} = \frac{14.605 - 0.860}{2}$$

$$= 6.87 \text{ in.} \quad (\text{try } 7 \text{ in.})$$

Try stiffener plates $\frac{3}{8} \times 4 \times 0$ ft-7 in.

$$A_{st} = 2(0.375)(4) = 3.0 \text{ in.}^2 > 1.52 \text{ in.}^2 \qquad \text{OK}$$

From AISC Specification 1.9.1.2, the width-to-thickness ratio

$$\frac{w}{t} = \frac{4.0}{0.375} = 10.7 \le \frac{95}{\sqrt{F_y}} = \frac{95}{\sqrt{36}} = 15.8 \qquad \text{OK}$$

Force on stiffener welds. Force at column web to be resisted by stiffener welds

$$P_{bf} - F_{yc}t(t_b + 5k) = 213.8 - 36(0.525)[0.595 + 5(1.5625)] = 54.9 \text{ k}$$

and the force at the tension flange to be resisted by the stiffener welds

$$P_{bf} = \frac{t_{fc}^2 f_{yc}}{0.16} = 213.8 - \frac{(0.860)^2 36}{0.16} = 47.4 \text{ k}$$

The column web force governs. The load on stiffener welds = 54.9/2 = 27.4 k each side.

Stiffener weld requirement

1. Weld size: From Table 8-4 (AISC Specification 1.17.2):
   Minimum weld size to column web is $\frac{1}{4}$-in.
   Minimum weld size to column flange is $\frac{5}{16}$ in.
2. Weld lengths:
   Length for a $\frac{1}{4}$-in. weld at the web:

$$l_w = \frac{27.4}{2(4)(0.93)(1.67)} = 2.05 \text{ in.}$$

Make a $\frac{1}{4}$-in. fillet weld full length of stiffener,

$$7 - 0.75 = 6.25 \text{ in. both sides}$$

$$(0.75 \text{ in. cut from corner of stiffeners})$$

Length for a $\frac{5}{16}$-in. weld at the flanges:

$$l_f = \frac{27.4}{2(5)(0.93)(1.67)} = 1.76 \text{ in.}$$

Make a $\frac{5}{16}$-in. weld full length of stiffener,

$$4.0 - 0.75 = 3.25 \text{ in. both sides}$$

Check of shear stress in weld metal

$$f_v = \frac{27.4}{2(\frac{5}{16})(0.707)(3.25)} = 19.1 \text{ ksi} < 0.3(70)(1.67) = 35.1 \text{ ksi} \qquad \text{OK}$$

Check of shear stress in stiffener base metal

$$f_v = \frac{27.4}{\frac{3}{8}(3.25)} = 22.5 \text{ ksi} < 0.4(36)(1.67) = 24.0 \text{ ksi} \qquad \text{OK}$$

(The allowable stresses were multiplied by 1.67 because the load on the stiffeners already includes a factor of safety.) The design is shown in Fig. 8-24.

PROBLEMS

Use the AISC *Specifications* for the following problems.

8-1. Two PL$\frac{3}{8}$ × 6 made of A36 steel are joined by a lap joint as shown in Fig. 8-6. Determine the required minimum lap and transverse weld size to develop the full tensile capacity of the plates.
(a) Use the SMAW process and $F_u = 70$ ksi for the weld metal.
(b) Use the SAW process and $F_u = 70$ ksi for the weld metal.

8-2. Repeat Prob. 8-1 with A572 Grade 60 steel with matching electrode metal (Table 8-1). Use the SAW process.

8-3. Two PLt × 8 of A36 steel are joined together by a groove weld. The plates support a tensile load of 85 k. Determine the plate thickness required and from Table 8-1 the matching flux electrode combination. Use the SAW process and complete penetration square-groove weld.

8-4. Repeat Prob. 8-3 using A572 Grade 50 steel if the tensile load is increased to 100 k.

8-5. A plate t × 6 of A36 steel is fillet welded to a gusset plate as shown. Determine the required plate thickness and weld lengths for SAW process welds if the weld size and flux electrode combination are as follows:
(a) Largest efficient weld size, F6X-EXXX.
(b) Smallest possible weld size, F6X-EXXX.
(Round plate thickness to the next highest $\frac{1}{16}$ in.)

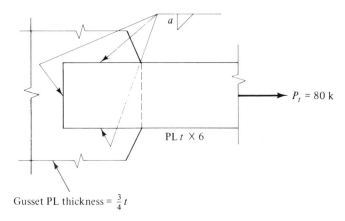

Gusset PL thickness = $\frac{3}{4} t$

Prob. 8-5

8-6. Repeat Prob. 8-5 using A572 Grade 50 Steel and F7X-EXXX flux electrode combination.

8-7. A two-angle l × l × $\frac{1}{2}$ tension member of A36 steel is fillet welded by the SAW process with $F_u = 60$ ksi for the weld metal to a gusset plate as shown. If the weld is balanced, determine the required angle size and weld lengths if weld sizes are as follows:
(a) $a = \frac{3}{8}$-in. welds.
(b) $a = \frac{5}{16}$-in. welds.

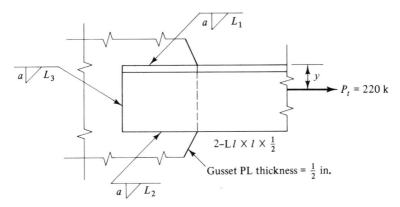

Prob. 8-7

8-8. Repeat Prob. 8-7 with $a = \frac{3}{8}$-in. welds and weld lengths as follows:
  (a) Use $L_1 = L_2$ (welds not balanced).
  (b) Omit the end weld ($L_3 = 0$).

8-9. Repeat Prob. 8-7 using A572 Grade 50 steel and F7X flux electrode combination.

8-10. An angle L8 × 6 × $\frac{1}{2}$ of A36 steel is fillet welded to a gusset plate as shown. Determine the tensile capacity of the angle and if the connection is balanced, the lengths of weld required if the weld size, electrode metal tensile strength, and welding process are as follows:
  (a) $a = \frac{7}{16}$ in., $F_u = 60$ ksi, SMAW process.
  (b) $a = \frac{3}{8}$ in., $F_u = 70$ ksi, SAW process.

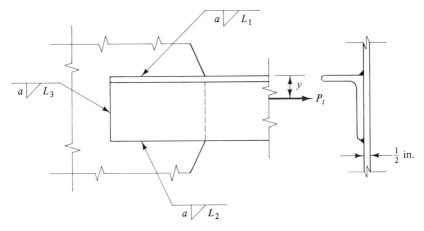

Prob. 8-10

8-11. Repeat Prob. 8-10 with L4 × $3\frac{1}{2}$ × $\frac{5}{16}$. The weld size, electrode metal tensile strength, and welding process are as follows:
  (a) $a = \frac{1}{4}$ in., $F_u = 60$ ksi, SMAW process.
  (b) $a = \frac{1}{4}$ in., $F_u = 70$ ksi, SAW process.

8-12. Repeat Prob. 8-10. Use A572 Grade 50 steel and electrode metal with a tensile strength $F_u = 70$ ksi.

8-13. Use the elastic method to determine the maximum force in k/in. on the weld group shown and the weld size required using the SAW process and weld metal with $F_u = 70$ ksi if the load P and lengths L_1 and L_2 are as follows:

    (a) $P = 50$ k, $L_1 = 10$ in., $L_2 = 4$ in.

    (b) $P = 40$ k, $L_1 = 9$ in., $L_2 = 3$ in.

    (c) $P = 35$ k, $L_1 = 8$ in., $L_2 = 2$ in.

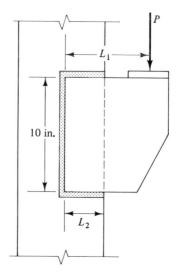

Prob. 8-13

8-14. Use the ultimate strength method (AISC *Manual*) to determine the weld size in Prob. 8-13.

8-15. Use the elastic method to determine the maximum allowable load P for the bracket shown based on the strength of the weld. Assume that the SMAW process was used with E70XX electrodes.

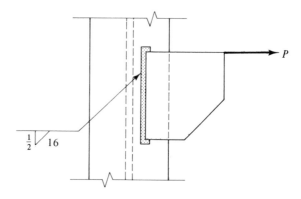

Prob. 8-15

8-16. Repeat Prob. 8-15 with the SAW process and $F_u = 70$ ksi for the weld metal.

8-17. Use the elastic method to determine the weld size required for the eccentrically loaded connection made of A572 Grade 50 steel as shown. Use matching electrode metal with
   (a) The SMAW process.
   (b) The SAW process.

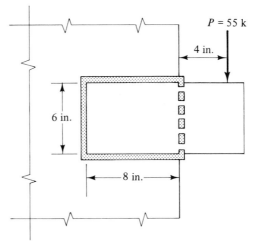

Prob. 8-17

8-18. Repeat Prob. 8-17 using the ultimate strength method (AISC *Manual*).

8-19. Repeat Prob. 8-18 with the eccentric load $P = 65$ k.

8-20. Use the ultimate strength method (AISC *Manual*) to determine the weld lengths L required to support the eccentric load shown when using the SMAW process

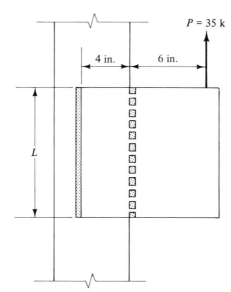

Prob. 8-20

with E70XX electrodes and weld sizes as follows:

(a) $\frac{5}{16}$ in.

(b) $\frac{3}{8}$ in.

8-21. A bracket is welded to a column as shown in Fig. 8-18. If $P = 20$ k, $e = 8$ in., $d = 10$ in., and the SMAW process is used with E70XX electrodes, determine the weld size required.

(a) Use the elastic method.

(b) Use the ultimate strength method (AISC *Manual*).

8-22. If $P = 35$ k, $e = 6$ in., and $d = 10$ in., determine the weld size for the connection shown in Fig. 8-18 if the SMAW process is used with E70 electrodes.

(a) Use the elastic method.

(b) Use the ultimate strength method (AISC *Manual*).

8-23. Design the welds and framing angles to connect a W21 × 68 beam to a W12 × 96 column. The sections are of A36 steel and the reaction on the end of the beam $R = 55$ k. Use the SMAW process and E70 electrodes.

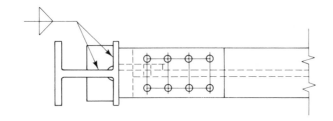

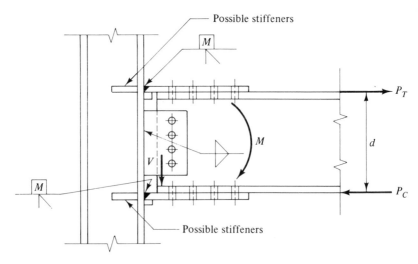

Moment connection with flange
and vertical web plate

Prob. 8-29

8-24. Design the welds and framing angles to connect a W21 × 50 beam to a W30 × 116 girder. The sections are A36 steel and the beam reaction is $R = 80$ k. Use the SMAW process, E70 electrodes, and the AISC *Manual* tables.

8-25. Design a seated connection to support a W16 × 31 beam of A36 steel. The reaction on the beam is $R = 28$ k. The beam is connected to a column web by welds. Determine the beam seat angle and weld size required if the SMAW process is used with E60 electrodes.

8-26. Design a seated connection to support a W18 × 71 beam of A36 steel. The reaction on the beam is $R = 32$ k. The beam seat is connected to a column flange by welds. Determine the beam seat angle and weld size required if the SMAW process is used with E70 electrodes.

8-27. Design a stiffened beam seat connection for a W24 × 84 beam of A36 steel. The reaction on the beam $R = 62$ k. The beam seat is connected to a column flange by welds. Use a structural tee section as shown in Fig. 8-22. Determine the weld size and structural tee required if the SMAW process is used with E70 electrodes.

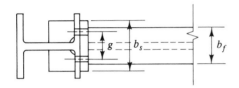

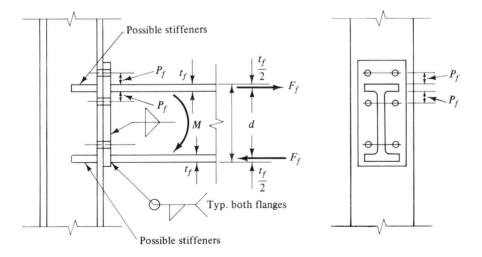

End plate moment connection

Prob. 8-31

8-28. Repeat Example 8-14 with a two-plate welded stiffener seat. Use the AISC *Manual* tables.

8-29. Design a moment connection with flange and vertical web plates to join a W21 × 50 beam to a W12 × 96 column as shown in the figure. The design moment $M = 100$ k-ft and the shear $V = 42$ k are the result of dead and live loads only. The members are made of A36 steel with $F_t = 22$ ksi (for beam $F_b = 24$ ksi). Use $\frac{3}{4}$-in. A325-N bolts and welds made by the SMAW process and E70 electrodes.

8-30. Repeat Prob. 8-29 if the moment connection is used to join a W21 × 44 beam and a W14 × 74 column, $M = 109$ k-ft, $V = 48$ k, and $\frac{7}{8}$-in. A325-N bolts are used.

8-31. Design an end plate moment connection to resist the maximum negative moment and end reaction for a W12 × 45 beam connected to a W12 × 96 column as shown in the figure. The members are of A36 steel. Use A325-N bolts and welds made by the SMAW process with E70 electrodes.

8-32. Repeat Prob. 8-31 if the connection is used to join a W14 × 43 beam and a W10 × 100 column.

INDEX

ANSWERS TO ODD-NUMBERED PROBLEMS

CHAPTER 2

2-1. (a) 67.2 psf
(b) 50.1 psf

2-3. (a) B_2: 87.5 psf, B_3: 86.8 psf, and B_4: 85.9 psf
(b) C_2 and C_4: 59.8 psf

2-5. $P_1 = 2.97$ k, $P_2 = 5.94$ k

2-7. $P_1 = P_7 = 1.33$ k,
$P_2 = \cdots = P_6 = 2.67$ k

2-9. *Side wall:* 3.00 k and 1.50 k
Roof: 3.37 k and 6.73 k suction

CHAPTER 3

3-1. 7.69 in.^{2}
3-3. 5.53 in.^{2}
3-5. 7.10 in.^{2}
3-7. 6.36 in.^{2}
3-9. 7.86 in.^{2}
3-11. (a) 9.46 in.^{2}
(b) 14.74 in.^{2}
3-13. (a) 60.5
(b) 69.3

(c) 87.2
(d) 163.1
(e) 83.3
3-15. 195 k
3-17. $\frac{5}{8}$ in.
3-19. Two L5 × 5 × $\frac{5}{16}$ or two L6 × $3\frac{1}{2}$ × $\frac{3}{8}$
3-21. Two L5 × 5 × $\frac{1}{2}$ or two L6 × 4 × $\frac{1}{2}$
3-23. L4 × 4 × $\frac{1}{4}$ or L5 × 3 × $\frac{1}{4}$
3-25. W10 × 45
3-27. WT8 × 25
3-29. C12 × 25
3-31. WT5 × 16.5 or WT7 × 17

CHAPTER 4

4-1. (a) 277 k
(b) 57.7 k
(c) 228 k
(d) 31.5 k
4-3. (a) 623 k
(b) 1046 k
(c) 759 k
(d) 667 k

(e) 308 k

(f) 214 k

4-5. (a) 376 k

(b) 75.8 k

(c) 113.6 k

(d) 180.7 k

4-7. (a) 74.9 k

(b) 109.6 k

(c) 205 k

(d) 111.5 k

4-9. (a) 298 k

(b) 420 k

(c) 576 k

(d) 686 k

4-11. (a) 284 k

(b) 350 k

(c) 529 k

(d) 645 k

4-13. (a) 403 k

(b) 466 k

(c) 539 k

(d) 588 k

4-15. W12 × 87

4-17. (a) W12 × 79

(b) W12 × 87

(c) W12 × 96

(d) W12 × 106

4-19. (a) Two L4 × 3 × $\frac{1}{4}$

or two L3$\frac{1}{2}$ × 3$\frac{1}{2}$ × $\frac{1}{4}$

(b) Two L4 × 3$\frac{1}{2}$ × $\frac{3}{8}$

or two L4 × 4 × $\frac{3}{8}$

(c) Two L6 × 3$\frac{1}{2}$ × $\frac{3}{8}$

or two L5 × 5 × $\frac{3}{8}$

(d) Two L6 × 4 × $\frac{1}{2}$

or two L5 × 5 × $\frac{1}{2}$

4-21. (a) WT7 × 19

(b) WT5 × 15

(c) WT6 × 20

(d) WT6 × 13

4-23. Four L5 × 5 × $\frac{3}{4}$, h = 23 in.

4-25. Two C10 × 20

4-27. (a) 382 k

(b) *Single lacing:*
$\frac{7}{16}$ × 2$\frac{1}{2}$ × 1 ft-7$\frac{1}{2}$ in.
Tie plate: $\frac{5}{16}$ × 14 × 1 ft-7$\frac{1}{2}$ in.

4-29. (a) 526 k

(b) *Single lacing:* $\frac{1}{4}$ × 2$\frac{1}{2}$ × 11$\frac{1}{2}$
Tie plate: $\frac{3}{16}$ × 7$\frac{1}{2}$ × 11$\frac{1}{2}$

4-31. (a) 440 k

(b) 454 k

(c) 355 k

(d) 396 k

4-33. (a) PL1$\frac{3}{8}$ × 14 × 1 ft-4 in.

(b) PL1$\frac{5}{8}$ × 16 × 1 ft-6 in.

(c) PL1$\frac{13}{16}$ × 19 × 1 ft-9 in.

(d) PL2$\frac{1}{2}$ × 17 × 1 ft-8 in.

CHAPTER 5

5-1. M10 × 9

5-3. W24 × 55

5-5. W27 × 94

5-7. W27 × 94

5-9. W24 × 94

5-11. W18 × 50

5-13. (a) B_1: W10 × 22, G_1: W16 × 40

(b) B_1: W12 × 22, G_1: W21 × 44

(c) B_1: W16 × 31, G_1: W21 × 68

(d) B_1: W18 × 40, G_1: W27 × 84

5-15. (a) B_1: W14 × 22, G_1: W16 × 40

(b) B_1: W12 × 22, G_1: W21 × 44

(c) B_1: W16 × 31, G_1: W21 × 68

(d) B_1: W18 × 40, G_1: W27 × 84

5-17. (a) B_1: W14 × 22, B_2: W12 × 14
B_3: W14 × 22, B_4: W10 × 12 or
M12 × 11.8, B_5: W8 × 10, G_1:
W18 × 35, G_2: W18 × 35

(b) B_1: W14 × 22, B_2: W12 × 14
B_3: W14 × 22, B_4: M12 × 11.8
B_5: M10 × 9, G_1: W18 × 35
G_2: W18 × 35

(c) B_1: W18 × 40, B_2: W12 × 26
B_3: W18 × 40, B_4: W12 × 19
B_5: W12 × 16, G_1: W24 × 62
G_2: W24 × 62

(d) B_1: W18 × 50, B_2: W16 × 31
B_3: W18 × 50, B_4: W14 × 22
B_5: M14 × 18, G_1: W24 × 76
G_2: W24 × 76

5-19. (a) L_1: 24 ksi, 534 k-ft
$\quad\quad L_2$: 22 ksi, 490 k-ft
$\quad\quad L_3$: 20.4 ksi, 455 k-ft
(b) L_1: 24 ksi, 392 k-ft
$\quad\quad L_2$: 22 ksi, 359 k-ft
$\quad\quad L_3$: 18.15 ksi, 296 k-ft
(c) L_1: 24 ksi, 163.2 k-ft
$\quad\quad L_2$: 22 ksi, 149.6 k-ft
$\quad\quad L_3$: 16.87 ksi, 114.7 k-ft
(d) L_1: 24 ksi, 292 k-ft
$\quad\quad L_2$: 22 ksi, 268 k-ft
$\quad\quad L_3$: 20.6 ksi, 250 k-ft

5-21. (a) 15.8 k
(b) 29.2 k
(c) 61.1 k
(d) 152.0 k

5-23. (a) PL1 × 10 × 1 ft-2 in.
(b) PL1$\frac{1}{2}$ × 12 × 1 ft-7 in.
(c) PL2 × 12 × 1 ft-11 in.
(d) PL1$\frac{3}{8}$ × 10 × 1 ft-3 in.

5-25. (a) W21 × 122
(b) W18 × 97
(c) W27 × 161
(d) W27 × 178

5-27. (a) W30 × 173
(b) W24 × 131 or W21 × 132
(c) W24 × 104
(d) W24 × 162

5-29. (a) W27 × 94
(b) W21 × 111
(c) W27 × 94
(d) W27 × 94

5-31. (a) W24 × 94
(b) W18 × 76 or W24 × 76

5-33. (a) W18 × 60
(b) W12 × 58

5-35. (a) W24 × 84
(b) W24 × 68 or W21 × 68

CHAPTER 6

6-1. Yes, interaction equations equal to 0.953 and 0.653.

6-3. No, interaction equations equal to 1.048 and 0.915.

6-5. Yes, interaction equations equal to 0.877 and 0.758.

6-7. W8 × 58

6-9. W12 × 72

6-11. W14 × 120

6-13. W12 × 45

6-15. (a) W14 × 61
(b) W14 × 68

6-17. WT8 × 22.5

6-19. W14 × 82 slight overstress or W14 × 90

6-21. Two L6 × 6 × $\frac{3}{4}$

6-23. WT10.5 × 55.5

6-25. $K_x = 1.92$, W14 × 159

6-27. $K_x = 1.18$, W12 × 106

CHAPTER 7

7-1. (a) 94.5 k
(b) 94.5k
(c) 101.5 k
(d) 101.5 k
$\quad l_e = 1\frac{1}{2}$ in., $s = 2\frac{5}{8}$ for all parts

7-3. (a) 18 bolts, 9 on each end of the splice
(b) 12 bolts, 6 on each end of the splice
(c) 8 bolts, 4 on each end of the splice
(d) 8 bolts, 4 on each end of the splice

7-5. (a) 14 bolts, $l_e = 1\frac{1}{2}$ in., $s = 2\frac{5}{8}$ in.
(b) 8 bolts, $l_e = 2$ in., $s = 2\frac{5}{8}$ in.
(c) 10 bolts, $l_e = 1\frac{1}{4}$ in., $s = 2\frac{1}{4}$ in.
(d) 8 bolts, $l_e = 1\frac{1}{4}$ in., $s = 2\frac{1}{4}$ in.

7-7. (a) 14 bolts, use 16 bolts
(b) 9 bolts, use 10 or 12 bolts
(c) 12 bolts
(d) 7 bolts, use 8 bolts
$\quad l_e = 1\frac{1}{4}$ in., $s = 2\frac{1}{4}$ in. for all parts except **(d)** $l_e = 1\frac{1}{2}$ in.

7-9. (a) 11.94 k
(b) 9.56 k

7-11. (a) 16 bolts, 4 in each row
(b) 20 bolts, 5 in each row
(c) 20 bolts, 5 in each row
(d) 24 bolts, 6 in each row

7-13. (a) connection is adequate
(b) connection is adequate

7-15. (a) A: 5 bolts, $s = 2\frac{5}{8}$ in., $l_e = 1\frac{1}{2}$ in.
B: 8 bolts, $s = 2\frac{5}{8}$ in., $l_e = 1\frac{1}{2}$ in.
(b) A: 4 bolts, $s = 2\frac{5}{8}$ in., $l_e = 1\frac{7}{8}$ in.
B: 6 bolts, $s = 2\frac{5}{8}$ in., $l_e = 1\frac{1}{2}$ in.
(c) A: 5 bolts, $s = 2\frac{1}{4}$ in., $l_e = 1\frac{1}{4}$ in.
B: 8 bolts, $s = 2\frac{1}{4}$ in., $l_e = 1\frac{1}{4}$ in.
(d) A: 4 bolts, $s = 2\frac{1}{4}$ in., $l_e = 1\frac{3}{8}$ in.
B: 6 bolts, $s = 2\frac{1}{4}$ in., $l_e = 1\frac{1}{4}$ in.

7-17. Capacity $R = 84.2$ k for $s = 3$ in.,
$l_e = 1\frac{1}{4}$ in., and $l_h = 1\frac{3}{4}$ in.

7-19. Use 4 bolts for web connection, 8
bolts for flange connection, $\frac{5}{16}$-in.
web angles, 3-in. bolt spacing, and
edge distances $l_e = 1\frac{1}{4}$ in. and $l_h = 1\frac{3}{4}$ in.

7-21. Use 5 bolts for beam web connection, 10 bolts for girder web connection, $\frac{5}{16}$-in. web angle, 3-in. bolt
spacing, and edge distances $l_v = 1\frac{1}{4}$ in. and $l_h = 2\frac{1}{4}$ in.

7-23. Yes, with $\frac{3}{8}$-in. web angle, 3-in. bolt
spacing, and edge distances $l_e = 1\frac{1}{4}$ in. and $l_h = 1\frac{1}{2}$ in.

7-25. Use 3 bolts and seat angle
L4 × 4 × $\frac{3}{4}$ × 0 ft-8 in.

7-27. Use 4 bolts and seat angle
L6 × 4 × $\frac{3}{4}$ × 0 ft-8 in. (4-in. outstanding leg)

7-29. Use 6 bolts in three rows.
Seat angle: L5 × 5 × $\frac{3}{8}$ × 0 ft-8 in.
Stiffeners:
Two L4 × 4 × $\frac{5}{16}$ × 0 ft-10$\frac{1}{8}$ in.
Filler plate: PL $\frac{3}{8}$ × 5$\frac{1}{2}$ × 0 ft-8 in.
Top angle: L4 × 4 × $\frac{1}{4}$ × 0 ft-8 in.

7-31. *Bolts:* $\frac{3}{4}$-in. diameter A325-N.
Web angles:
Two L5 × 4 × $\frac{5}{16}$ × 0 ft-5$\frac{1}{2}$ in. with 2
beam web bolts and 4 column
flange bolts.

Tees: Two WT12 × 38 × 0 ft-8$\frac{1}{2}$ in.
with 4 tee flange bolts and 6 tee
stem bolts.
No stiffeners are required.

CHAPTER 8

8-1. Minimum lap = 1$\frac{7}{8}$ in.
(a) $\frac{5}{16}$-in. fillet welds
(b) $\frac{1}{4}$-in. fillet welds

8-3. PL $\frac{1}{2}$-in. thick, F6X-EXXX
flux electrode combination,
prequalified square-groove
weld designated B-L1-S.

8-5. PL $\frac{5}{8}$-in. and gusset plate $\frac{1}{2}$-in. thick
(a) $\frac{7}{16}$-in. weld 12 in. long
(b) $\frac{1}{4}$-in. weld 18 in. long

8-7. Two L6 × 6 × $\frac{1}{2}$
(a) $L_1 = 9$ in., $L_2 = 2$ in.,
$L_3 = 6$ in.
(b) $L_1 = 12$ in., $L_2 = 3$ in.,
$L_3 = 6$ in.

8-9. Two L4 × 4 × $\frac{1}{2}$
(a) $L_1 = 8$ in., $L_2 = 3$ in.,
$L_3 = 4$ in.
(b) $L_1 = 10$ in., $L_2 = 3$ in.,
$L_3 = 4$ in.

8-11. $P_t = 49.5$ k
(a) $L_1 = 9$ in., $L_2 = 3$ in.,
$L_3 = 4$ in.
(b) $L_1 = 5$ in., $L_2 = 1$ in.,
$L_3 = 4$ in.

8-13. (a) 10.34 k/in., $\frac{9}{16}$-in. weld
(b) 9.90 k/in., $\frac{9}{16}$-in. weld
(c) 7.40 k/in., $\frac{3}{8}$-in. weld

8-15. P = 29.8 k

8-17. (a) $\frac{7}{16}$-in. weld
(b) $\frac{5}{16}$-in. weld

8-19. (a) $\frac{1}{2}$-in. weld
(b) $\frac{3}{8}$-in. weld

8-21. (a) $\frac{3}{8}$-in. weld
(b) $\frac{1}{4}$-in. weld

8-23. Two L3 × 3 × $\frac{1}{2}$ × 0 ft-9 in.,
$F_y = 36$ ksi

Weld A: $a = \frac{1}{4}$ in., SMAW process, E70 electrode

Weld B: $a = \frac{3}{8}$ in., SMAW process, E70 electrode

8-25. One L7 × 4 × 1 × ft-8 in. with 4 in. outstanding leg and $\frac{5}{16}$-in. welds with $\frac{1}{2}$-in. returns at the top of the welds.

8-27. Use WT13.5 × 51 × 0 ft-5 in. with $\frac{1}{4}$-in. weld (SMAW process, E70 electrode).

8-29. *Bolts:* $\frac{3}{4}$-in.-diameter A325-N bolts. *Flange plate:* $\frac{1}{2}$ × 6 in. top and bottom with 8 bolts for the flange connection.

Web plate: $\frac{1}{4}$ × 5 × 15 bolted to the beam web with 5 bolts and welded to the column flange with $\frac{5}{16}$-in. fillet welds full length both sides.

8-31. *End plate:* $\frac{7}{8}$ × $9\frac{1}{2}$ × 1 ft-3 in. welded to the flanges of the beam by $\frac{1}{2}$-in. fillet welds and the web of the beam by $\frac{5}{16}$-in. fillet welds. The end plate is attached to the flange of the column with 4 bolts at the tension flange of the beam and 2 bolts at the compression flange (1-in.-diameter A325-N bolts).

Stiffener plates:

Four $\frac{3}{8}$ × 4 × 0 ft-6 in. plates welded to the column flange by $\frac{5}{8}$-in. fillet welds and the column web by $\frac{1}{4}$-in. fillet welds.